ALBATROSS

BIRDS OF FLIGHT - BOOK ONE

J. M. ERICKSON

outskirtspress
DENVER, COLORADO

This is a work of fiction. The events and characters described herein are imaginary and are not intended to refer to specific places or living persons. The opinions expressed in this manuscript are solely the opinions of the author and do not represent the opinions or thoughts of the publisher. The author has represented and warranted full ownership and/or legal right to publish all the materials in this book.

Albatross
Birds of Flight - Book One (Revised)
All Rights Reserved.
Copyright © 2014 J. M. Erickson
v2.0

Editor: Suzanne M. Owen

Cover Image by Joseph M. Ericson

This book may not be reproduced, transmitted, or stored in whole or in part by any means, including graphic, electronic, or mechanical without the express written consent of the publisher except in the case of brief quotations embodied in critical articles and reviews.

Outskirts Press, Inc.
http://www.outskirtspress.com

ISBN: 978-1-4787-3079-8

Library of Congress Control Number: 2014903171

Outskirts Press and the "OP" logo are trademarks belonging to Outskirts Press, Inc.

PRINTED IN THE UNITED STATES OF AMERICA

Praise for *Albatross: Birds of Flight*

"...Erickson rolls out his narrative with no-nonsense storytelling...It's easy to get caught up in the story...because we trust the characters..." *-Blue Ink Review*

"...This is a thinking man's thriller, filled with moral dilemmas and chessboard strategy. I thoroughly enjoyed it and would not hesitate to recommend it..." *-Readers' Favorite*

"...Erickson competently portrays the sometimes violent tactics, though it's clear that his real interest lies in his characters' emotions and psychology..." *-Kirkus Review*

"...Combining insights into criminal psychology with his knowledge of law enforcement, he has penned a realistic thriller replete with unexpected twists and cathartic turns..."- *IP Book Reviewers*

"...J. M. Erickson elevates the premise beyond the Bourne-again shtick with an interesting twist and taut, engaging writing that will keep even jaded readers hooked..." *-Blue Ink Review*

"...Apart from being a thriller, this novel is also a story of betrayal, romance and redemption...Above all, it is a moving story about a man's personal journey and redemption..."- *Readers' Favorite*

"...Erickson doesn't sacrifice explosions and gunfire; instead, he peppers the action with psychological insight..." -*Kirkus Review*

"...in addition to a great political thriller, there are also underlying themes of romance and redemption. These factors all add up to a very compelling, fast paced adventure..."- *Reader Views*

Note from the Editor

Albatross is merely the first chapter in the story of government special agent Alexander Burns and his mission to achieve justice and regain both his lost memory and his sense of self. Along the way, he is supported by a team of disparate individuals who are drawn together by the mission itself as well as by their developing relationships and loyalties.

One of the greatest pleasures for me in editing this novel was observing the growth of and interaction between the various characters. They struggle throughout to figure out their roles and their own purposes while the subplots intertwine and take shape, all of which is set off by the direct, clean writing. The language also shifts effectively between various registers, from colloquial speech by the characters and narrator to military terminology as well as the refined language of literary references, and this further enhances the atmosphere of the scenes and the psychological details at play.

My objective in editing this novel was to ensure the consistent "pull" of the plot and subplots and the seamlessness of the shifts in language. I was also mindful of maintaining the directness of the writing, especially the narration, so that one feels present in the story at all times, observing the action while being able to reflect on the changes in the characters and plot over time. The smoothness and pace of the language move the reader through the novel as plot details continually emerge, drawing the reader to explore the next complex, thrilling chapter.

Suzanne M. Owen, Editor

Contents

Prologue	i
Chapter 1	1
Chapter 2	27
Chapter 3	35
Chapter 4	44
Chapter 5	67
Chapter 6	86
Chapter 7	99
Chapter 8	121
Chapter 9	127
Chapter 10	150
Chapter 11	177
Chapter 12	192
Chapter 13	203
Chapter 14	230
Chapter 15	252
Chapter 16	270
Chapter 17	281

Chapter 18	319
Chapter 19	340
Chapter 20	352
Chapter 21	364
Epilogue	389
List of Characters	393

O happy living things! no tongue
Their beauty might declare:
A spring of love gushed from my heart,
And I blessed them unaware:
Sure my kind saint took pity on me,
And I blessed them unaware.
The selfsame moment I could pray;
And from my neck so free
The Albatross fell off, and sank
Like lead into the sea.

– Samuel T. Coleridge,
The Rime of the Ancient Mariner, IV. 65–6

Prologue

"Clara pacta, boni amici"
"Clear agreements, good friends"

Five Years Earlier – May 1

"BURNS! WHAT IS YOUR problem? We have an opportunity to capture the second most dangerous asshole in the world, and you want to bag and tag him? For a logistics specialist you're pretty short sighted," Maxwell said as he burst into his small, unairconditioned room.

Closing his eyes and opening them again, Burns focused on his laptop screen as he responded.

I just don't have time for this.

"Look...the President authorized both my plan and the Seal's plan to kill our respective assholes. The Seals will get Bin Laden tomorrow and we'll cap our boy tonight. Why aren't you crawling up their ass about capturing their target?"

"Because they're stupid. I'd expect that patriotic bullshit from them. But you, Burns? An opportunity to capture this guy and seize his entire network? I can't believe you proposed this plan against mine? And Daniels is pretty pissed with you floating your idea to the Navy on this. He doesn't like going outside the chain of command...Burns? What the hell are you doing?" Maxwell asked as he came around to see what he was doing.

You have no idea what I've been up to.

Clad in his military issued fatigues, Burns flexed his fingers and stretched his back to help organize his thoughts. After shifting just a little bit in his leather back chair, just to remove the stickiness of his sweat from his clothes, he tried to assess what to tell his colleague and what to leave out.

"After twelve straight hours, I found a very simple, basic design flaw in our computer servers' network firewall," Burns said calmly as he closed the top secret file containing the possible locations of where his agency would move to, should their location be compromised. He did, however, leave another top secret file, *Albatross*, open for Maxwell to see just to prove that he had access to top secret data.

"What the hell is *Albatross*?" Maxwell asked as he was scanning the document over Burns's shoulder.

"It's a very nasty Trojan virus, developed by the NSA and perfected by our guys. A nice little computer virus that is designated for testing on the North Koreans in three months," he responded calmly.

"Jesus, Burns. What is your problem? Why are you always trying to prove yourself? Why are you always trying to second guess Daniels? It only pisses him off," Maxwell said as he stood back up and started to walk out.

"And Burns? Get your gear – we're on deck in one hour; briefing in two, and prep to leave for launch point in Bravo

Base near the Swat Valley, Pakistan in three. We'll be testing out new attack-stealth choppers for the first time in the field so we should get there early for a dry run. That's unless you decided to join our Seal friends instead," he added sarcastically while he continued walking.

Looking at Maxwell as he retreated, Burns found himself speaking before he knew it.

"We should kill Sudani. We don't need him and his network. He killed our people. Our soldiers. He and his friend brought the fight to our home. He's an asshole and deserves nothing more than a swift execution so we can go on with our lives."

Burns watched Maxwell come to an abrupt halt and then turn around to look at him.

Shock? That's the look. I know. I don't know where this tirade is coming from either.

"Burns...what's going on with you?" Maxwell asked, genuinely concerned.

Looking down at his keys, he thought about it for a moment before he spoke.

"I'm tired. I'm tired of this 'War on Terrorism.' It's time we bag and tag these leaders, these terrorists. End the terrorism, the loss of life, innocent civilians here and there. It can only end with their deaths. Not negotiation with them in exchange for their minions. How are we different from them if we allow them to just live in peace while our guys continue to get killed? You know what will happen if we capture Sudani – Daniels will strike a deal with him, hide him, and then take out the superficial guys while the real shit keeps going. More death. No end. I'm just tired of the shit.

Aren't you tired, Maxwell? Don't you just get tired of all the bullshit?" he asked, as a wave a fatigue came over him even as he sat in his chair.

Burns watched Maxwell's eyes narrow as he stood speechless.

"You need help Burns," he finally said and walked out of the room.

Alone, Burns looked back down at his computer screen and turned it off only to catch his reflection in the dark monitor. Looking blankly at his face, he could easily see the bags under his eyes, and more wrinkles around his mouth as he listened to the room's silence.

"Well...that didn't go well," Burns said to himself.

24 hours later – May 2

"...we're in the kill zone! Those missiles aren't coming from the surface - they're coming from above us! What the hell is going on? Our target has no air cover?" the pilot said as she fought to keep the helicopter in the air. Ignoring the obvious, Burns spoke calmly as he watched the streaks of missiles heading right towards him. Shooting off flares to obscure the missile's guidance systems, he knew it was only a matter of time before he and the small command helicopter would be hit. And still, he felt unusually calm, almost as if being shot down or worse was not as horrible as it could get.

I should have seen this one coming!

"Maxwell – Call off your birds. This is *Falcon 5*. I need to confirm the target's body."

"Burns?! You're breaking protocol! I got this – don't worry about it and turn back before you get blown out of the sky!"

I'm breaking protocol? How about you follow my directions! Focus, Burns...

"Nothing personal but I want to make sure we got the right target and that there were no changes in plans," he said calmly as a way to get Maxwell talking. Feeling the maneuvering helicopter banking a hard right while his stomach churned upside down, Burns covered his microphone to listen to what the pilot was saying.

"Sorry – we're in way over our heads. I'm turning back."

"No problem. I'll keep trying to get them to back off..."

"Don't they know we're on the same team?" she said with anger as she began a rapid ascent to avoid yet more weapon fire.

Yeah...they know we're on the same team, except I'm expendable now, he thought as he rapidly replayed the sudden changes in aircraft assignments that kept him in the small, unarmed scout helicopter that was in "command position," way back from the field of battle to quarterback the mission. Not responding to her question, Burns focused on his control panel to field the flares, radar and talk to Maxwell.

"Maxwell – what were your orders? What was the encrypted message from Daniels? I know it came in...I saw it come over the private channel" he said as he launched yet more flares to confuse the incoming missiles as the helicopter descended suddenly but still under the pilot's control. With proximity alarms falling silent briefly, he focused as best he could to hear what his "friend" was saying.

"How? How did you..." was all Burns heard before a missile alarm blared out again but this time, rather than it turning off after a series of death-defying acrobatics, there was a hard jolt in the rear, causing a loud grinding metal sound and engine strain.

"Shit! Rear rotor's gone – no pressure back there...half of my controls are down," she said. Watching his entire panel flicker, spark and go dark, a strong smell of electrical wire burning wafted in from below his feet and behind his seat. Dark black smoke began to rapidly fill the cockpit as Burns simply listened to anybody who would talk to him.

"Mayday, mayday, mayday – Command Helicopter 525 hit from friendly fire...We are going down near grid 626 – DG. We're going down hard," she uttered out as Burns shifted his focus beyond her and saw the ground fast approaching.

"Sorry Alex," he heard both clearly and unexpectedly in his headset from Maxwell. Hearing the same thing, the pilot, a middle-aged woman with blonde hair and safety glasses, turned to look at him, acknowledging that she caught the not-so-hidden message. Feeling his stomach rise to his throat, the acidic, electrical smell filling his nostrils, and heat emanating from his damaged panel, he made very brief eye contact with the pilot through the smoke.

"That shithead set us up. Nice friend," she said as she returned to raising the nose of the helicopter to get it to crash land on its skids rather than immediately kill them in a head-on collision into the ground.

"I don't have any friends," Burns said with more resignation than he expected.

"Twenty feet," she said. Feeling the helicopter stabilizing and the cockpit leveling off, it was easy to feel that they were coming in way too fast. A sudden image of a woman, Debbie Foley, came to his mind as he closed his eyes, waiting for the inevitable crash.

She shouldn't have gone to Heathrow – she was a logistics specialist, not a field agent...what was Daniels thinking, he wondered as he heard the pilot talk for what he assumed would be the last time.

"Ten feet...damn it! This is going to be hard on my kids," she said with a sad calm.

Great...another person's life I've screwed up..., he thought as he felt his body begin to press into his seat, restrained by his safety harness, while sounds of metal snapping, crunching and buckling filled his ears. While time passed slowly, so did the experience of sharp shards of metal and glass piercing his skin, but nothing in comparison to his hands and arms that felt as if they were on fire. When blocking out the pain was no longer possible, he felt more shards explode near his head, leaving his hearing tinny and shooting pain burning his scalp. After some time, he felt as if finally the world stopped spinning and tumbling, leaving the strong smell of burning wires. While in the absence of sounds, he took in a deep smoke, acid breath as a way to gear up to opening his eyes, but coughed, as he began feeling lightheaded and warm liquid on his head. Still though, with his arms and hands on fire, he tried opening his eyes, but all he saw was a bright, white light that was closing down fast as blackness encroached rapidly.

So...this is how this life ends...

Twelve minutes later...

Watching his field agents methodically fan out around the still smoldering crash site, Anthony Maxwell was still listening to his boss's ridiculous recommendations as he scanned the horizon looking for telltale road dust from either the local authorities or Pakistani troops.

It's not going to take the locals or the curious long to find this, and we're not supposed to be here, he thought as he turned his gaze to two newer field agents removing a dead pilot, a woman, killed on impact. Shifting his focus to the passenger side of the helicopter, it was easy to see a blood trail at first but preliminary reports had it running dry after thirty feet. Looking to the horizon again, he watched as three helicopters flew low in a clear search pattern. After a few moments, Maxwell realized that his immediate boss, Thomas Webber, was now suggesting that he contact the American Embassy to get troops to assist in the search. Feeling his neck muscles tighten and his eyes enlarge, he now focused on the satellite radio and spoke at it as if Webber were right in front of him.

"Are you crazy? We're not supposed to be here! How the hell will this be a covert operation if we mobilize the marines to assist us find one of our guys when we aren't supposed to be here, and we have no jurisdiction or authority to conduct any operation here? How about you leave me alone and I'll give you a situation report in twenty," Maxwell said as he closed out the channel without waiting for a response.

"Jesus..." was all he was able to utter when he turned to see one of the newer agents waiting for him to finish telling him something.

"What? What is it?" Maxwell said as he lifted his binoculars to scan the horizon again. Hearing the pale young man clear his throat, Maxwell was already planning to talk to the clerk who assigned two newbies to such a classified assignment.

"Sir, why do we have to bring a body back? I mean, he's clearly injured and if the heat and desert doesn't get him, robbers, the enemy or animals will," he said. Taking a deep breath, Maxwell tried to remember that unlike Webber, this guy was just learning.

OK...just take it slow and don't kill him. The asshole who made the roster will pay for this shit decision.

"Obviously you have no idea whom you're dealing with. Alexander J. Burns not only is an experienced field agent but a tactical specialist who knows more secrets than he should. And bringing back a body to show the White House will confirm he was killed in action..."

"So like I said, just let the desert do its job," the young agent said before Maxwell finished. Turning on him, Maxwell felt like striking the man but decided to focus on the recovery operation and finding the right words. Maxwell also had a curious thought as to how much longer a pale man like the agent in front of him would last in the desert as he watched him liberally apply suntan lotion.

"First – don't ever interrupt me when I'm speaking, dumb-ass. Secondly, with Burns missing in action, I have no proof that he's dead. And unlike you...whatever your name is, Burns has skills in surviving. If he's out there and alive, that means he'll be back and pissed off. It's not going to take him long to figure out that we screwed him. So how about you focus on your job and let the grownups do the thinking," Maxwell said as he returned to scanning the horizon. Without looking back at the agent, he sensed him moving off quickly as he saw one of the helicopters breaking off its search pattern and move directly towards him.

Please let this be good news, he hoped as his operations radio chirped on.

"Recon 6 to CO: we've got two dust trails coming in – one from the east and the other the south. ETA is ten minutes. If we want to keep this on the down low, we need to abort the search and evac immediately. Copy?" Looking to his east and then his south, he could barely make out the dust trails the pilot had seen.

"Damn it...all right, Recon 6, bring everyone in and let's get out of here. Have Team 12 land and demo the crash site. Get a message to command, and skip Webber and go directly to Daniels on my authority. Message is as follows: 'falcon has left falconer; his condition unknown.' Got it?"

"'Falcon has left falconer; his condition unknown.' Got it. Recon 6 out."

Still scanning the horizon, Maxwell watched the two other helicopters break off their search and return to the crash site while the field agents on the ground also returned quickly to the landing zone. As sun rise was only an hour old, Maxwell felt uneasy not knowing whether his former friend and colleague was dead or alive.

"Damn it Burns! Why couldn't you just go with the program?"

Chapter 1

"Fallaces sunt rerum species"
"The appearances of things are deceptive,"

– Seneca

Present Day – May 2

WHERE AM I? I was on my way to the car and then...
Anthony Maxwell was just waking up when he felt a burning sensation in his arm and a headache forming in the back of his head. He started to move his limbs, but they were firmly bound to a chair. Maxwell's eyes were having difficulty adjusting to the room as a result of bright lights shining on him. He had been in the intelligence business long enough to know that he was being interrogated. While he had witnessed many interrogations and conducted a few himself, he had never been the subject of one.
Shit. This is bad. Real bad.
Though he was not fully conscious, he could sense someone was in front of him, sitting and waiting. He was sure the burning sensation in his right arm was an intravenous

concoction to make him talk. His mouth was dry, and he wanted to talk; however, his entire body seemed to be devoid of liquids, especially saliva, which made talking very difficult. Maxwell knew he had to collect his wits and try to remember how he ended up in this terrible predicament. He remembered walking to his car after his impromptu meeting with his two contractors to provide a final briefing on a "foreign agent" who was living right over the border in Canada. It was a small mission of information gathering, and the briefing was supposed to finalize the logistics.

How could they get the drop on me? What am I? A recruit?!

Focusing on breathing while drawing moisture from his throat so he could talk, Maxwell pulled himself together as his eyes adjusted and he familiarized himself with the stark room. In general, he was all about security and being careful; being a senior field agent of the Department of Defense Foreign Intelligence's Operations Center always meant being vigilant. If you wanted to live, the practice of being vigilant was a lifestyle and not just a good habit.

So whoever was able to track me and get the drop on me before I was able to discharge my weapon was either very lucky or a professional ... or both.

With more saliva forming in his mouth, he was better able to croak out a question.

"Do you know who I am?"

"Yes," said a soothing, low, and calming voice.

OK. Well...under vastly different circumstances, this voice could have be comforting. Warm and inviting even, he thought.

"What do you want?" Maxwell went on. From years of training and experience, it was easy for him to tell that he was in a room with wooden walls and windows.

Maybe a house. Neighborhood? Abandoned area? If I scream, I probably could be heard, provided I'm still in a populated area. It feels like wood is under my feet.

While it took time for his eyes to adjust, Maxwell could tell there was someone behind his interrogator. The man behind his interrogator seemed to stand still and watch the entire interaction.

Something about the way the guy is standing seems familiar, Maxwell thought to himself.

"I don't want anything. We have what we want. Now we wait," the voice went on. Maxwell shifted his focus back on the interrogator sitting right in front of him.

"Look. I'm a senior field agent of the Department of Defense. If I don't check in with my people in a couple of hours, the federal government will be looking for me. That means they will be looking for you. Do you get it?" Maxwell said in an attempt to tip the tables to get his interrogator talking. He was hoping for the usual bravado, machismo, or arrogant response to his threat. Though he knew he posed no real danger while he was tied up in a chair, he still wanted to get a dialogue going. What he got was more chilling than he wanted to admit.

As the silence continued, Maxwell wondered if maybe his captors were rethinking their situation.

Maybe I can get my ass out of this situation, he thought.

Then the interrogator gave a sigh, which was followed by an even-toned reply.

"You are Anthony R. Maxwell. You are a senior field agent of the Department of Defense's Foreign Intelligence

Agency assigned to the operations center located in Waltham, Massachusetts. Just so you are aware, the medication flowing into your arm is not any drug that will make you talk. It is a combination of Vicodin and Valium that will relax you and allow you to nod off and fall asleep," the voice articulated.

All right. This is very bad. This guy has my real name, the location of the operations center, and has no interest in information. Damn it! That means these two already have what they want, making me disposable. It's time for a new strategy.

"Well, maybe you might want to know some classified data? Do you have any idea how much I know and what you could get for it?"

Time to make some offers, bargain, and buy some time, he thought as he felt his head cloud over and a yawn breaking through as he struggled to stay awake. The more he struggled, the more tired and sleepy he got as if his moving and talking had drained all his energy.

What is this shit they're giving me? It's gotta be more than just Vicodin and Valium.

It was hard to think. Maxwell was getting sleepy. His arms, legs, and stomach started feeling like lead. He had to focus.

"No, Mr. Maxwell. There is no need for that. We already texted your contact team to meet you here in three hours. They will more likely be here in two hours or so," the interrogator concluded. But then there was a follow-up question.

"Mr. Maxwell, you do know what today's date is, don't you?"

What? Odd question, Maxwell thought. *Maybe there is something about today's date that is either an anniversary or a target or mission date.*

"May 1. 'May Day' in Catholic tradition. It's an important

day for the old Communist—"

I'm so tired...why is the date so important? What do you want?

Maxwell attempted to keep talking, but he was fading much faster than he thought. It was difficult for him to form thoughts, let alone sentences.

"No, Mr. Maxwell. Today's date is May 2," the voice corrected. Maxwell took note of the tone: there was no judgment in the interrogator's voice; it was just a correction of the date.

I can't think...I can't stay awake. So tired. My head is so heavy, Maxwell thought as he felt his head drop and then jerk up in an attempt to stay awake.

Got to focus...

Looking up Maxwell saw the shadow behind the interrogator move slowly toward him as he was fading. As he was nodding off, he knew from the person's build and profile that there was something definitely familiar about it.

Maybe it's somebody from my past. Someone I pissed off or worse. North Korea? No – Russia...The date is so familiar.

Maxwell began to feel light-headed and elated. Then, before he completely slipped away, he uttered what he hoped was an understandable sentence. Maxwell was sure it would be the last set of cogent questions he would ever ask.

"Burns? It's you? You're MIA? What the hell..."

It was easy for him to see the shadow stop moving while the interrogator crossed his legs. Maxwell looked away for a moment to try to focus on something else to stay awake. His mind was wandering. While his first thoughts were on old friends and family, they faded too quickly, he thought. Instead of seeing other friendly faces, he saw the faces of past enemies, victims, and collateral damage he had caused. He couldn't help but feel weighed down by these thoughts. He shook his head to clear his thoughts, but the faces stayed in view.

Hallucinations? he questioned.

"Why them? Where's Daniels? Foley? My parents?" he asked out loud.

Maxwell looked back at the interrogator and then at the shadow, and he was positive the man was Alex Burns — one of the faces he was seeing.

"Why him?" he tried to finish.

But Maxwell reached that critical threshold where he no longer cared about the world anymore. All his cares seemed to recede. The faces were the last to fade. He drifted off into an opiate-driven sleep.

The shadow waited just outside the rear of the house, where he had passively watched Maxwell's interrogation hours before. With full knowledge that his prisoner was in a deep sleep, still bound to the chair in the middle of an empty house, the shadow recalled the last time he and Maxwell actually worked together.

So many years ago. A lifetime ago. I'm not even the same man anymore. Still...Let's just hope your guys don't get jittery and start shooting in there, he thought.

As part of their business, they never used first names. First names were too personal. Last names only were used both in the field and off. Seeing "Maxwell" in his drug-induced condition did make Burns feel bad. After so many years of his being missing, he was amazed that Maxwell did recognize him as "Burns." He was also amazed that he felt no enjoyment at Maxwell's situation. In the past, he would have had no empathy or sympathy for his victim. Even though he had good reason to hate Maxwell, he felt bad for him. Burns's companion, the even-spoken interrogator, was not surprised that Burns had empathy for Maxwell. As Burns waited outside the house, he smiled at the interaction he had with the interrogator, David Caulfield, right after Maxwell had passed out. David had predicted that he would feel bad for him.

"You know, David, for a trained mental health professional, you're not very good at hiding your emotions from your clients, especially when you're right. Very good interrogation too, I might add," Burns commented.

"Well, you're not my client anymore. Not for some time."

He had watched David interrogate Maxwell and was impressed with David's natural ability to be soothing even in such a terrifying situation. Still, Burns did find it unnerving that David could guess how he felt pain for Maxwell. In the past, Burns was unreadable to colleagues, superiors, and enemies. He could not remember having friends or close family.

Am I that readable? Poker face gone? Maybe they could read me back then too and I was just too focused on the missions to notice, he thought.

"It's not my first day on the job. Interviewing, that is. It also helps not being able to see what's really happening. I can pretend I am in an office," David continued.

Burns knew David would need assistance to reach the basement. Once they navigated through the rooms and down the flight of stairs where David would wait, he said to Burns,

"Please be careful."

"Will do. You too. It might get noisy upstairs so stay down here," Burns told him as he guided David into a waiting chair before he left him to wait outside in back of the house.

An integral part of plan required that David would have to be "found" in the basement by the authorities. After he was finished setting the stage upstairs, after the other cast of characters arrived, he would have to assist David 'get into character.' Burns was not looking forward to that part. He decided to try to recall more positive thoughts for a moment. It was easier now in the last four years; he had positive thoughts and memories to now draw from. His friendship with David was one of them. He was still smiling and moving his feet to stay warm when he had the urge to scratch his scalp where old scars prevented hair from growing. The scars on his hands and arms, however, would itch because of dryness.

Dryness? No – it's nerves and anxiety. Admit it. Welcome to the living, Burns.

As always, his thoughts focused back on the plan, and he began yet another process of reviewing possible scenarios. If all went well, Maxwell's contacts will arrive first, and the FBI agents next. It was very dark, and the evergreen trees offered excellent cover. Though he would have preferred a moonless night rather than the new moon in the sky, he was at the mercy of nature if he wanted to make sure everything happened today — not the day before or tomorrow, but today.

Well, good news here – it's not raining. And there's not another soul in sight. It's just a matter of time. No turning back now.

Burns chose this location and this house because it was

the only one that was close to being completed and ready for occupancy. He also chose the house because it was located near the woods, which gave him a perfect line of sight from the back door as well as a view of the driveway in the front of the house.

For the moment, Maxwell's car was the only one visible in the driveway. After two hours, a second set of headlights drove down the road. Before it turned into the driveway, the headlights turned off and the car blocked Maxwell's car.

Here we go.

Two occupants exited their car and approached the house as quietly as possible. Burns watched them enter the house cautiously at first — one in the front and the other in back. It remained quiet inside and outside of the house until yet another set of headlights pulled up behind the two parked cars.

Perfect. Now all you guys have to do is just sit still for a minute, he thought.

Winter was over and the sun rose in New England at 5:20 a.m. It was by no means sunlit at 6:20 a.m. inside the house, but it wasn't pitch dark either. The two newer arrivals slowly exited their car, each looking at the house and assessing their next move. This pair had an air of "law enforcement" about them; they stood at an angle to the house, making themselves less of a target. Their hands were firmly placed on their hip holsters, where Burns was sure they each carried a .9mm semiautomatic weapon. One of the agents was decidedly taller and had a lanky build, while his partner was of average height but clearly stockier.

As the occupants of the last car were closing in on the front of the house, Burns saw the first pair exiting the back as quietly as possible.

Nope. That's not part of the plan. This is the part where you hold ground.

As the back door opened, Burns steadied his stance and carefully aimed his own semiautomatic to the left of the doorframe. The crack of the gun report was loud in the suburban neighborhood of empty houses. The two who were exiting the house now backed away rapidly as a second report shattered wood on the right side of the doorframe. The two men in front crouched slightly and produced their own weapons as they approached the front of the house. Burns emerged from the woods and circled wide around the house, keeping his eyes on the living room windows.

"FBI! Come out with your hands up!"

Not original, but it's clear.

Suddenly, there was yelling from inside the house: "You set us up, asshole!" There was a single shot.

Ah, shit! Maxwell.

Burns heard the front door break open. Shouting and yelling erupted inside the house, and the shouts were confusing to understand.

All right. Time to end this.

Burns steadied his weapon and took four shots towards the pair he had kept at bay in the house. He planned on giving the federal agents an edge. Shots fired in all directions from inside the house and then there was silence. Burns stood for a moment to make sure there was no movement in the house. Normally, he would have taken the time to collect his shell casings so that he could eliminate any connection between his gun and the crime scene. If he really wanted to clean up the crime scene, he would have to eliminate his footprints, dig out the slugs that had to be lodged into a wall or ceiling, and wipe down all of his fingerprints inside the house. However, Burns wanted to make sure there was no confusion that his bullets were not involved with any deaths in the house and wanted a "big X" to show everyone he was outside when his gun discharged. That was important because he wanted to make sure he was in the clear; he did not want to be seen as a killer.

Not me. Not today, at least.

Burns then quietly walked to Maxwell's car, opened the trunk, and took out a full paramedic case. He took off his black jacket, which concealed the standard paramilitary white shirt with epaulets, and neatly placed his jacket in the trunk. He then took out and put on the standard, bright orange first-responder jacket with reflective stripes, and literally changed persona from "shadow" to "paramedic." Even before he entered the building, he applied his latex-free gloves and holstered his semiautomatic gun, which was concealed inside the jacket.

Burns carefully entered the house and stated loudly and firmly "I'm a paramedic. I am unarmed and coming in through the front door. Don't shoot."

Let's hope someone is left alive to help, let alone respond.

As Burns came through the door, he could smell the carbon of recently discharged weapons. Opening the front door, he implemented two of the three rules in first aid — survey the area and provide care. The third and final rule, "call for assistance," he planned to do much later than a real paramedic would in such a situation. The first federal agent, the stocky one, dropped his gun as soon as he saw the "paramedic." The agent then moved his empty hand back to his left shoulder so he could continue compressing his own wound. It was also clear his right thigh had been hit too. Burns squatted beside the stocky agent, opened his kit, and took out dressings and bandages. He provided first aid, and at the same time, he kept monitoring if there was any movement in the other rooms.

As the agent looked on quietly, he asked about his partner.

"I don't know yet. I'll finish with you and get to him," Burns responded.

"Don't you have another guy? You guys travel in pairs, right?" The agent asked.

Yup. Real paramedics do. Former counterterrorist specialists don't.

"Usually, but we are really short staffed and there's a lot of activity tonight. Be happy I was on break down the street when the call came in," Burns lied.

Well, let's hope he buys that bull.

Offering no more questions, the agent sat quietly and allowed treatment to proceed without another word. Once first aid was completed, Burns moved to the tall, lanky agent who was lying face up with a shot in his chest and a smaller injury to his ankle. Fortunately, Burns had been recently briefed on assessing such a wound. First, he made sure there were signs of life, which miraculously there were. The agent was still breathing. Burns immediately opened the agent's jacket, applied pressure, and tried his best to dress the wound. Once he was satisfied the compression had slowed the bleeding, he wanted to avoid more serious injuries caused by the lanky agent falling into shock by dragging him to the relatively "healthier" stocky agent. As soon as Burns had the lanky agent lying next to his partner, he started an intravenous line.

"Now you need to hold this above your friend's head so he gets the fluid. If you don't, he will die," Burns warned.

And that will keep you focused on just him and nothing else.

He gave the stocky agent back his and his partner's guns and walked to where the other victims were. There were three. Maxwell was still bound to the chair but lying on his side, motionless. Burns was caught by surprise when he experienced regret that Maxwell was now dead.

Damn it. That's just not right. Your guys are real assholes.

He knew it would take the authorities time to identify Maxwell because Burns had taken all of Maxwell's identification cards and corresponding badges. The other two dead bodies he didn't need to know about.

Without even pausing to look at them, Burns continued walking through the living room to the kitchen and went downstairs to the basement. With the lights off. Burns couldn't remember if he had left them off when he had brought David down or not.

You know. A simple detail and you missed it. Focus, Burns.

He hesitated and then turned the lights on, and with his own weapon drawn, he descended quietly down the stairs. As he turned the corner, he saw that David was sitting in nearly the same position he had originally left him in hours prior.

Wow. What is wrong with this picture? he thought as he shortened the gap between them.

His friend looked out of place. As if a mid-forties man in a navy blue suit, matching tie, and a stark, white shirt sitting in the dark below a crime scene wasn't odd enough, the fact that David was wearing dark sunglasses in the dark basement was unnerving.

"I hate that you wait in the dark like that with your sunglasses on," Burns said and smiled.

"I have no need for light," David responded.

Calming voice as always, he thought.

Because both men had interrogated Maxwell and the stage was now set, David stood up, moved the chair away from him, and handed Burns, his former client, his sunglasses. For whatever reason, Burns's hand missed the glasses. Without showing annoyance, David bent over to feel for them as Burns struck the back of David's head. David slumped downward toward the hard cement floor and would have smacked his face with force if he had not been caught in time by Burns.

"I'm sorry," Burns said to his now-unconscious friend. He really hated this part of the plan. Allowing David to be taken into custody wasn't going to be any easier, either.

After more than four years of dealing with counterintuitive feelings, conflicting thoughts, and diametrically opposing behaviors, Burns was getting used to his complex identity. He laid his former therapist on the floor and exited the basement through the bulkhead. He turned to the driveway and got behind the wheel of the federal agents' car after he retrieved a laptop from Maxwell's car. As he started the car, he took out a cell phone. It was a cheap, prepaid cell phone and was very hard to trace. Its clock showed 6:35 a.m.

Well, that was a fast fifteen minutes.

As there were no sounds of first responders, he called 911 and reported that he thought he had heard men shouting and gunshots at the new Leveritt development. Almost on cue, he heard sirens as the operator informed him that help was on the way. Other residents must have heard the shots too. Burns hung up the phone and backed the FBI agents' car out of the driveway, leaving the other two cars in place.

A half of a mile away at a traffic light, he rapidly started typing on a separate smartphone: "Black knight in place. White knight on the move. Alpha out."

Funny. What if I got pulled over for texting while driving? he thought as he pressed "send."

This smart phone was paid through a cell phone company, which did make it possible to track. But in this case, that was all part of the plan.

Even though Burns had meticulously planned the next several steps, any mission was vulnerable to human or technological errors, second thoughts, and direct interventions from the government. Maxwell's death was a clear example of not being "part of the plan," even though the risk was accounted for. While Burns felt confident in his own abilities, he was nervous about his team.

They're just civilians caught up in some serious shit, he thought.

Burns found himself drifting off of the mission again. It was a common problem now. In the past, he had laser focus during an operation. Now he had worries about whether his friends were up for this major undertaking or not. He was especially worried about the woman he loved. Again, without his full memory of his past, the absence of the memory of loving someone was telling. That made Samantha the only woman he had ever remembered loving. As Burns drove, he smiled. He always smiled when he thought of her.

Samantha was sitting at the nursing station when she had gotten the text. She had been waiting for some kind of text, positive or otherwise, since the beginning of the third shift. Because Laura, one of the first-shift nurses, had called in sick, she was doing a double. As the first- and third-shift doctors, nurses, and medical technicians were now in rounds, reviewing last night's events and scheduling out the new day's appointments, Samantha had volunteered to cover the front desk and patiently wait for this very text. She looked carefully at it to make sure it was not the abort code.

Well. It starts, she thought.

Convinced the text said what it meant, she placed her smartphone deep in her pocket and picked up the nursing station line and put the empty phone line on hold. After she placed the phone back in the cradle, she then went to find Jack, the first-shift security officer who would more likely be in the front lobby of the hospital's main entrance. The main entrance lobby of Saint Lawrence Hospital was positioned to look out over the old section of Lawrence. Across the street were the old mill buildings that the city was now converting into multi-income condominiums. The condominiums were situated to look over the Merrimack River. The glassed-in lobby gave a panoramic view of the city.

"Jack?" Samantha asked as she approached the security officer. While she was new to the staff, Samantha soon learned from them that Jack was a legend among the female nurses. While she could see elements of how he might have been quite the catch in the past, his thinning hair, thick middle, and tight clothes were detractors. She had to give him credit though.

You still think you're God's gift to women, she thought.

"What is it, Ms. Smith?" Jack replied. Samantha watched him look her up and down, smiling as usual.

Could you be less obvious about it, or is subtlety not your strong suit? she thought.

She ignored him. He always checked her out, and today, she made sure to have her makeup just right. Her hair was its customary raven black and shoulder length, and her uniform seemed a bit tighter than usual. None of this was really needed to get his attention.

You know, I could wear a barrel or a potato sack and I bet big money you'd check me out anyway. You have no idea how much you'd have to pay for me, she mused.

"Jack," she repeated, "I have a guy on the phone who says there is a bomb in the emergency room nursing station."

Jack's smiled faded a little bit but returned quickly.

"Well, did you see one? You were just there," he said.

Oh my God. You think if I saw one I'd be telling you about some guy on the phone talking about it? OK – focus on being pretty and dumb, she reminded herself.

"I didn't look. The guy told me to go out into the lobby and look outside to make sure I believed him," she offered.

Samantha watched Jack stand up from his chair behind the podium desk and look out the lobby windows. He was about to say something when a flash of red, white, and yellow flames shot out of the top-floor windows in one of the condo units facing the hospital. The sound finally caught up with the sight, and the windows shook violently.

Samantha stepped back, and Jack turned ash white.

Wow! How the hell did he get that to be so big?!

Even though she expected an explosion, Samantha was caught off guard at the size and violence of it. Once she regrouped, she saw Jack run back to call for police and noticed that one of the lines was lit with someone on hold. He looked at her, and she immediately picked up the phone.

In a shaking voice, she finally asked, "Did you set that off?"

She waited a moment. She then responded to a fictional voice by saying, "What do you mean ... there's one in the second lateral file near the nursing station in the ER and one on the maternity floor?"

Samantha made sure to widen her eyes as if in shock or surprised.

You have no idea how many times I've played that role for men. Men are so easy!

As the fire grew, consuming the buildings, a number of staff came out to look at the ensuing flames. Cell phones were ringing, and the switchboard was beginning to light up. Jack stayed transfixed on the nurse, waiting for more information.

Distracted by the commotion, Samantha returned to her role of talking to the fictitious caller.

"Hello? Hello? Are you there?"

Jack snatched the phone from Samantha and listened, but there was only silence. Jack hung up the phone and pulled her with him to the back doors. While Samantha and Jack headed back to the emergency room, they were slowed by the tide of staff heading toward the lobby to see what had happened. In full knowledge of what to expect, she came around to the other side of the nursing station and looked for a lateral file cabinet. Samantha realized that for the first time, Jack seemed more interested in his job than her backside. Right now, he was in full security mode and was the first to see a lateral file at the far end of the desk.

I am impressed, Jack. You're taking your job seriously. Nice to see.

Watching him approach the cabinet slowly and begin to open it, the she whispered, "Be careful, Jack."

If she had not been there with her hand gently touching his back, she was not sure he would have opened the file cabinet at all.

I know if it were me, I'd be gone.

She could see that machismo got the better of him. As Samantha watched Jack open the cabinet, she saw blinking red, yellow, and green electronic light-emitting diode lights with wires going in and out of four separately wrapped blocks.

Well, that does look ominous. I'd definitely run if I saw that.

There also seemed to be a white piece of paper with some writing on it. The writing seemed foreign with the exception of the last part, which clearly read, "May 2." Jack stepped back, and Samantha audibly gasped.

"Isn't it May 2 today?" Samantha said quietly as if the sound could detonate the explosive.

I said that well. Not too "frightened" but not bland or casual, she thought.

It was easy to see that Jack was all business. She watched him move himself and her back to the main nursing station. To be convincing, Samantha allowed Jack to physically move her away from the dangerous-looking bomb. If it wasn't for the bomb threat and the impending crisis to come, Samantha might have chuckled that Jack finally got his wish to lay hands on her. But she was feeling some guilt, knowing that she was responsible for starting this part of the drama.

Jack turned suddenly to her and said, "Get an outside line and call 911. Get fire, police, and bomb squad to the hospital."

Samantha immediately went to the phone and started her tasks. At the same time, she heard Jack's voice going over the hospital's public announcement in a firm, calm fashion: "Code black. Code black. Maternity ward is the first to move, emergency room second, remaining floors next. Code black. Maternity room first and emergency room second. This is not a test. This is not a drill." With that public address, a flurry of activity began with medical staff going to all the rooms to round up patients for evacuation.

Good thing this all happened close to 7:00 a.m., when there are overlaps in both shifts.

After his announcement, Jack gave Samantha a reassuring nod and moved off to assist coordinating the hospital evacuation.

Samantha wrapped up her 911 calls to police and fire departments and, she added for good measure, a request for a hazardous material team.

During her calls, she was also multitasking on her smartphone. Her text was short and to the point: "Alpha, Charlie. White bishop on the move. Bravo out."

She pressed "send" and was on her feet with her jacket and her bag. As all the staff ran around her while she walked against the flow of people exiting the halls, she found what she was looking for and ducked into the nurses' changing room. It took less than five minutes to change. In no time, she removed her nurse uniform, black wig, and white shoes and replaced them with navy blue slacks, a white paramilitary shirt, and an orange safety jacket with EMT on the lapel and back. With her clothes changed, she was now in her "paramedic" identity. In addition to her change in uniform, she now emerged from the changing room with long red hair pinned up under a black cap with a medical insignia.

The former nurse "Smith" was now fully donned in a paramedic uniform, allowing her to completely blend into the chaos. While all medical staff and patients were evacuating to the parking lot, the designated "safe zone," Samantha headed into the "danger zone," where the bomb had been placed. Her destination was the ambulance parking lot, which was right next to the evacuating emergency room. After two attempts, she finally found an ambulance with the keys in the ignition. She got in, closed the door, turned the engine on, and started to carefully drive out of the parking lot, making sure to avoid hitting people passing and running away from the emergency room. At first, Samantha heard lots of sirens getting closer, and then they started to recede as she drove in the opposite direction.

Back on her phone, she typed another text: "White bishop secured transport. Black bishop, you are a go. Bravo out."

Samantha began to consciously focus on relaxing her arms and shoulders. She was so tense from the explosion, bomb, change of identities, and stealing an ambulance that her body felt locked up with stress.

I don't know how Burns did all this shit for a living without getting ulcers, she thought. Her mind stayed with Burns for a moment. He was probably the third person she had loved other than her sister, Becky, and her friend David. It was hard for her to sift out how she cared for David as opposed to Burns. She loved Burns, and their relationship was sexual; however, she never thought of him except by his last name — Burns. *Somehow, using the last name makes it safe for me to love him,* she thought.

She loved David but more as a father. The kind of father that treated children like children and not like miniature adults.

You know – real fathers that take care of their kids out of love.

Samantha shook off the memories of prior foster fathers and her own parents.

Pleasant thoughts, she told herself.

As for Becky, well, she was as close to family as Samantha ever knew. Of all the people she had known, she would put her life on the line to save her — no matter what.

As scared as she felt about what she was doing, she was doing it for Becky first and foremost.

As she drove toward the off-ramp, she wondered if Becky was faring better than she was in the espionage department.

What am I doing? I'm not a spy! I'm a mother now. I don't have time for this, she thought. *Still, if I don't, who will take care of Emma? The State? I don't think so.*

As Becky felt her phone vibrate, she completely focused her attention on the job at hand as she watched the third text come in. The first one gave her time to get the truck started. Prior to starting the truck, she flipped the door up on the flatbed after she made sure the barrels didn't move much when she drove. By the time the second text came in, she was making sure the menacing electronic box with blinking lights attached to the barrels was actually blinking. She had checked the batteries twice to make sure. When the third text came in, she was just covering the barrels and sides of the pickup truck with a heavy tarp, about to tie it down. She made sure to read the last text carefully. Once Becky finished tying the tarp up the sides of the flatbed truck, she climbed into the cab. She started the truck up, which stirred a sleeping four-and-a-half-year-old girl in the truck's passenger seat. The little blond-haired girl simply shifted her head to the other side of the seat to get comfortable, pressing her face into the seat even more.

She was glad that David wasn't there to see that Emma wasn't in her baby seat. He was really overprotective of her. He was overprotective of them both, actually. Becky smiled at the little girl again and then began the slowly drive to her athletic club.

You have no idea what we're all doing, Emma? It's a good thing. How the hell will I ever explain this when you're older? Hopefully I'll be around to.

She had been at this location more than a dozen times, so she knew the fastest way with fewest lights to get there. It was especially important to take the roads that had few or no traffic cameras. She was nervous when she heard a large number of sirens whirring in the distance. As she pulled the truck over the railroad tracks, she turned left into the industrial park where her health club was situated. She had chosen this particular athletic center because of its proximity to professional offices and the lack of surveillance cameras both in the parking lot and in the gym itself.

Jesus! For 7:20, the parking lot is pretty full.

While the other businesses were not open yet, the club was already abuzz with the overachieving moms, students, and professionals getting into their spinning classes and weight machines. Instead of going to her usual space closer to the gym, she parked right in the middle of the large parking lot away from the gym. Rather than parking within the parking lines, she parked across them and in the center of all the surrounding buildings.

Now that's really going to piss people off and get the much-needed attention I want, she thought as she smiled at the notion of obsessive-compulsive people angered by parking outside of the lines, taking up more than two car spaces.

"You see, I get that," she said quietly to herself as she unbuckled her belt. She hated it when people didn't park within the lines.

That's why we have lines – to stay between them to let others park. I hate it when people don't do that!

As casually as she could, she exited the truck, extracted a jogging stroller from the backseat, and placed her little girl in it. Becky had started running nearly three years ago when she finally had the motivation to stop smoking and stop eating everything in sight. Emma was one of those motivators. David, her boyfriend, was the other. Still though, even though she had the classic runner's body, she still would wear oversized tee shirts that covered her behind.

I should probably see someone about that.

After Becky made sure Emma was buckled into the jogging stroller, she put the keys in her running suit and pulled her dark brunette hair in under her hat. She then untied the tarps and flipped them into the truck, exposing writing on the side of the truck. The script was familiar to her — Arabic, she was sure. There was no mistaking the date stamped at the end of the writing: "May 2"! She also removed the tarp from off the top of barrels she had transported. As Becky came to the rear of the truck, she dropped the door and exposed what appeared to be some kind of control device with blinking lights, a device attached to taped blocks that had wires going into barrels in the flatbed. The sun was still below the tree line in the parking lot, but once it hit the tops of the barrels, it would ignite a dry-ice substance that would smoke profusely for about an hour until it dried out. That would draw a lot of attention, and she was sure to be long gone by then.

It won't be the smoke that gets people's attention in this town. Not parking within the lines will be enough to draw ire from this neighborhood, Becky thought.

She then casually put on her sunglasses and pulled out her smart phone and began typing as she used her stomach to slowly push the stroller away from the truck.

The text went out to two recipients: "Black bishop has delivered the package and is on the move. Heading to lair. Charlie out."

Becky pressed "send" and then placed her phone in her pocket.

"I hope all this works," Becky said to herself. She had noticed that since she had become a mom, she would often talk to herself. She thought it was because she was anxious. David told her it was to reassure herself. Still though, Becky never thought in a thousand years that she would be involved in such an elaborate plan. Years ago, planning her lunch at work was her big decision.

Second thoughts as always, she thought.

"Too late now," she muttered.

She pulled the oversized tee to cover her small frame and smoothed it out as she began her run back to the office. Becky did remember she had promised Emma a donut, so she knew she had to stop on the way. David would not be happy about the stop, but it was a rare treat. It had been a long time since she had had a donut. After she had lost nearly fifty pounds over the past few years, she had made it her life's work to reduce temptations.

But even David would say, "Once in a while is fine."

It's such a beautiful May day!

Chapter 2

"Crux"
"Puzzle"

Present Day – May 2

IT'S ALREADY A BAD day from the start, Andersen thought.

It began with a call at 6:00 a.m. from the watch commander about needing to redeploy the day shift to the neighboring town's hospital to deal with an actual bomb situation. There was also a burning building right across the street from the same hospital.

What are the odds of that happening? he thought.

To make matters worse, there was a shooting on the other side of his town in a new development. Sure, getting up early to get to work was really not that bad, but the day had proceeded downhill. He had cut himself shaving, dropped his coffee mug in the driveway, and dealt with more traffic than usual as he had headed to the richer side of North Reading. It was a pretty quiet town for a New England suburb of large homes with dual incomes. In these labile economic times, he was lucky to be employed.

Steve Andersen - lieutenant in the local police department, married to his wife for more than two decades and two healthy kids - considered himself lucky on many points. He even liked his in-laws. Even though his wife was sick with some stomach bug, at least she was not at the hospital where there was a bomb scare happening. She was pretty non-responsive when he told her about the hospital and the fire; she had been throwing up all night.

Man! What the hell were you doing last night? Drinking with the girls again? They're a pretty rough crowd and you're a lightweight, he had told her many times before when she attempted to keep up with her peers' bar antics.

It was still not a great day, though, to have multiple bodies in a new development of a new home in a soon-to-be secured, gated community.

That's going to really affect resale value if news of this spreads.

All these events fell in sharp contrast to a beautiful day. As Andersen approached Dempsey, the first officer to respond to the crime scene, he waved him into a driveway, where there was a large number of vehicles. Andersen could tell Dempsey was anxious from the way he waited for him to exit the car.

"So, what are we talking about?" Andersen quietly asked, cognizant of the various onlookers held back by yellow "Do Not Cross" tape.

"A lot of action for our neck of the woods," Dempsey replied just as quietly. He continued with his verbal report: "Three dead bodies total and two injuries. All white males in their late thirties to maybe early forties. One was bound to a chair and seems to have been executed. Shot in the back of the head at close range. He has no identification at all. The other two are heavy hitters from Boston. Drug traffickers, enforcers, the kind you would never expect here. Maybe South Boston, Dorchester, but not here. They were easy to ID as Organized crime."

South Boston? Really? Murphy and his crew maybe up here? North End and East Boston – the Panelli Family? What the hell are they doing business up here, Andersen thought as he took in the entire crime scene.

"That is weird," he said, more to himself than Officer Dempsey.

"It gets weirder," he continued. "The other two, both injured pretty bad but stable, the paramedics tell me, look like they just got out of Quantico — right down to black suits, shields, and guns."

"Jesus!"

Andersen knew he would have an hour at best before the FBI Boston's regional office took over his crime scene. Once the bureau showed up, Andersen would give in on the jurisdiction and hand the scene over to them. If it had been his guys who had been injured, he would have wanted the same professional courtesy.

Everyone on his team would have to move faster if they were going to get anything of value to solve this crime.

Dempsey continued with his report, "Yeah. And one other thing, there is a sixth guy. He was unconscious on the floor of the basement while everyone else was upstairs in a firefight. It's weird because he is well-dressed, well-groomed, and so out of place there that we thought he was a bank assessor for the house in the wrong place at the wrong time.

He is either black or Latino, in his late forties, no weapons, no cell phone. His ID puts him as a resident of Lawrence, and he has an office there too."

"What does he do?" Andersen asked finding himself feeling quite perplexed.

Three dead bodies in my town, one mystery man, two federal agents heading to the hospital in Wakefield because the local one is under threat, and there are no answers.

Looking back from where he came, he saw a news team setting up.

Of course, they're already on the scene. Makes sense, he thought. *It's big news in a quiet town.*

One ambulance was already gone with one of the federal officers; one pulled in pretty fast.

Everybody had to be short-staffed with evacuating the hospital, Andersen thought.

"That's the weird part," Dempsey continued. "He's a counselor of some sort."

"Are you kidding me?"

"Yeah. Think it was an anger management group that got out of control?" Dempsey queried through thin lips, suppressing a smile.

Andersen looked at him for a moment before he smiled.

Not bad, Dempsey. Not a bad joke. You're more of a slapstick guy and your delivery was really good. Maybe you do have gray matter up there, Andersen thought as he slowed his pace down to look at the spacing of the other homes.

At that moment, the paramedics were transporting the more seriously injured agent out of the house. It was a rather steep set of stairs but manageable. Andersen knew from his wife's nursing training and experience that in times of emergencies, medical staff would triage the least injured who could be saved first and would later get to the more seriously injured who were doubtful to survive. With the hospital's evacuation and the fire across the street, Andersen guessed that triage would be playing a major role for the first responders today.

In passing, the last paramedic pushing the stretcher down the walk said to Dempsey, "You guys did a great job getting a line in him. You might have saved this guy's life. No bruising too."

Dempsey looked confused and then turned to Andersen and gave him the I-have-no-idea-what-is-he-talking-about look. He then shrugged and confirmed his look by saying, "I have no idea what he means."

Well, maybe I jumped too soon at the thought of him having gray matter.

Andersen often wondered how Dempsey had gotten his position on the force — *an uncle or something*. Nepotism still played a role in getting civil servant positions.

OK. I need to focus on the case and not him. We got a live witness, an active crime scene and maybe an hour before the FBI show up. No pressure, Andersen thought.

"All right. Where is the guy now?" he asked as he surveyed the main room. There were bodies in various positions on the floor with spent shells. The smell of discharged weapons still hung in the air, and the room was cold, even though it was filled with the crime scene team. It didn't take extensive forensics training and experience to guess what had happened. The hostage had been shot in the back of the head and now he was on his side, chair forward. The two bad guys were behind him and presumably, one of the shooters had killed the guy in the chair while the FBI guys had killed them. Something else caught Andersen's eye though, and the crime scene guys were already all over it. There appeared to be bullet holes in the wall at an angle that could have hit the bad guys as well. On the other side of the wall, there was a set of windows that had glass blown inward. There was already someone taking casts of footprints outside.

There was someone else here. Another shooter. Someone shot at the bad guys, from what it looks like. So now we have a missing seventh person too. Just great.

One of the things that was also interesting was graffiti on what might have been the living room wall. Elegant shapes of script were clearly some form of Arabic writing. Already, there were staff members with their phones and digital cameras taking various angles of this too. The disturbing part was the English words that indicated today's date, May 2.

Hmm. Something about that date...

"So what about my witness?" Andersen came back to the present.

"Paramedics checked him out. There was a cut and a bruise on the back of his head, but that was it. He's on his way to the station now."

"All right. Get more details on who these guys are, extend the perimeter to the adjacent houses, and canvass the neighbors. Maybe someone saw something more. We have at best an hour before the Feds take over our crime scene. I'm going to talk to our number-one witness."

Dempsey smiled. "Well, he ain't talking much. He asked if anyone was alive upstairs, but that was it. He gave his name, occupation, where he lived, but that was it. Good news is he is not asking for a lawyer. One other thing: the guy is blind."

"No. Are you shitting me?

Of course he's blind! An eyewitness that's blind. Maybe he was just a bank assessor in the wrong place at the wrong time.

"No shit. Paramedics checked him out when they do their light-in-the-eyes thing, and there was no response."

"Great ... we'll be getting a call from the civil liberty and disabilities' advocates any minute now. The guy won't need a lawyer. Is he a veteran too?"

It was evident to Andersen that Dempsey sometimes couldn't tell if he was kidding or really asking. In moments like these, Dempsey would just remain quiet. But today was different. Dempsey had more to offer as he remembered a detail.

"Oh, yeah ... he asked to call his assistant so she could bring him some medication or something," Dempsey recalled.

Andersen always hated it when Dempsey would forget something like that.

"Did he make that call?" Andersen asked.

"One of the paramedics let him call on his phone. I think with all the craziness happening, the medics forgot he was in a crime scene," Dempsey finally concluded.

Maybe the day isn't so bad, Andersen thought. He had a live witness, no lawyers, and he would be the first to interview the witness.

Well, it's not all bad. It's time to get some answers.

"Call Jackson and Shelley and make sure they put him in

interrogation room eight. Use the VIP entrance out back so reporters and the public don't see our guy. No one in or out to talk to him. Give him water, but that's it. If his assistant gets to the station, have him or her wait. Once this area is cleared up, get to wherever they took our live agents and get some statements. So what's this guy's name?" Andersen asked as an afterthought as he entered his car.

"Samuel T. Coleridge," Dempsey answered.

Andersen stopped as he was about to get into his car and turned to Dempsey.

No! That's just not possible! What's the odds...

"Are you kidding? Wasn't he a poet? My witness is named after a dead poet," Andersen asked as he peered right through Dempsey to see if maybe the entire situation was some kind of elaborate joke.

"No idea," Dempsey responded.

Andersen wondered why he had even asked Dempsey that question.

"Things are pretty weird...a dead poet," he said more to himself than to Dempsey.

As he pulled out of the cluttered driveway, Andersen took in the view of all the official cars and the two cars of the victims inside. It took a moment, but then he noticed that neither car had government plates. He had to make a note to make sure that both plates were run. *Maybe the agents had been undercover or at least their ride had been,* Andersen thought.

Chapter 3

"Omnium rerum principia parva sunt"
"Everything has a small beginning,"

– Cicero

Present Day – May 2

ANDERSEN HAD BEEN AT the station for at least an hour before he got ready to enter the interrogation room. He had prepped the command officer and day-shift supervisor to make sure they could field the reporters' questions. He was also getting a steady flow of information about the victims. But why they were all in that one house at that time and what really had happened was still a mystery.

Andersen picked up a pad of paper, his favorite black pen, and his folder of reports and preliminary data. Rather than lingering in his office or going right to interview room eight, he took a moment to go to the bathroom. He stood in front of the mirror with the water running to get warm.

The hospital, condo fire, dead bodies, and federal agents shot up? This is something more, Andersen thought.

He would often sit in his office to prep himself for an

interrogation. But washing his hands and face seemed to relax him for this interview.

The guy's name is of a dead poet, and he's the only link to what is going on around here. I got to work this guy very carefully, he was reasoning.

Andersen lost track of what he was doing and had dried his hands very thoroughly with sandpaper-like paper towels. He adjusted his collar and checked his fly before he collected his material. Checking his fly was second nature.

How can you take an interrogator seriously if his fly is open? You only have to have that happen once to learn that hard lesson, he thought as he remembered his tour of duty at Guantanamo.

As he walked to the room, he thought he would call his wife. She was pretty sick and actually called in sick to work when he saw her last.

She has to be sick to actually call in. She never does that.

He was about to hit the speed dial but realized she would be sleeping, and he didn't want to disturb her. He decided to let her sleep. He was just glad she was not at work. *Fortuitous* was the best word he could think of for her absence at work that day.

Alright. I need answers, and this Coleridge guy is the guy to give me some.

Andersen had the dubious honor of being an urban legend in the world of interrogation in the thriving metropolis of North Reading, a suburb north of Boston. While it was true that his time questioning "combatants" at Guantanamo Bay was critical to expanding his skill set, he always felt it was luck and legend that would typically break the accused perpetrator or witness. Though during the war, Andersen's role in intelligence gathering was critical to finding top-tier assets, he often saw himself as "just a guy doing a job."

I might have missed nailing Bin Laden but I did get that other asshole, he often thought.

Still, he had regrets about his work back then. While there were indeed some bad guys back there, there were also some prisoners at Guantanamo that he genuinely thought played a minimal role, if any, as enemies to America. He tried to not let it bother him, but it did. Innocent people being locked up ate at him. That kind of knowledge was a burden. After a while, he had difficulty sleeping, stopped eating, and spent a great deal of time alone in his quarters. It took a couple of calls with his wife and his childhood friend, Diane Welch, to help him figure out what he had to do. He had to leave Guantanamo. At first, his superiors weren't going to let him go, but the medical doctors warned that continued weight loss and recurring fevers were not going to be good for him — or for the army's image if he died because of starvation. He was transferred to Germany in the role of critical incident specialist.

Laura and Diane saved my life, Andersen thought whenever he recounted his early days in the service.

Still, his time in Guantanamo was a key to his ethics around interrogation.

It's about the truth. If they're innocent, they walked out. If they're guilty, I lock them up. If this guy Coleridge is innocent, he'll walk, Andersen thought to himself.

Andersen put his hand on the doorknob, pushed, and entered the room.

Hopefully, he didn't hear me sigh. Still...Something's not right with this whole thing.

Interview room eight was twenty by twenty feet in size. The small room was lit by bright fluorescent lights and contained a heavy wood table and two old chairs. The glass mirror allowed viewers and cameras to watch from the other side while the person interrogated could see nothing but his or her own reflection.

As Andersen entered the room, he was struck by Coleridge's appearance. The man was well-dressed in a dark suit. Maybe it was black or navy blue. He always had a problem figuring out those dark colors. Fortunately, his wife always helped him out with that one. His wife could make everything match.

She was sophistication while he was bluntness. She read books while he preferred documentaries. She was also the holder of all his secrets. His siblings, best friend, and colleagues had no idea he loved classical poetry. She did. Laura was also a great sounding board for when he had difficult cases. Andersen was guessing this one might be one of them.

I hope she's feeling better, he thought.

The suit his witness had on looked like it had been made for a funeral.

At least it's not your own. Are you part of this shit or just lucky to be alive? What's your story?

His witness sported short black hair with lots of evidence of gray showing. Under the suit, the man wore a white collared shirt that had dirt or dust on it, and he was wearing dark sunglasses. Andersen could not see the matching tie.

Maybe he removed it. I'll have to check on that later.

Coleridge sat quietly at the desk as if he were waiting for a cup of coffee. To say that Coleridge was "calm" would be an understatement. While his light brown, clean-shaven complexion seemed youthful, with only a few creases betraying his age, Coleridge appeared to be a solidly built, athletic man.

Maybe a former football player, Andersen thought.

He did not seem to be your typical counselor. But then there were the scars that peppered the witness's face. Scars that were concentrated near the eyes and nose as well as a scattered pattern down his neck. Andersen was sure that if Coleridge's suit and shirt were removed, there would have been more scars visible, deeper ones as well. The scars were a lighter color of skin that had healed over, not a clear scar line one would have expected on a face to conceal an injury. In short, the man had never gotten plastic surgery to cover those scars.

So what type of counselor are you? Where did you get those scars from? How long ago?

And while Andersen had his own stereotypes of people with disabilities, the man in front of him seemed almost at home in the middle of the shit coming down around him.

Andersen quietly sat opposite of Coleridge. He opened his file, clicked his ballpoint pen for writing, and placed it carefully on the pad of paper. Andersen watched Coleridge as he sat impassively, as if waiting for his second cup of coffee. Andersen could see he was taking in everything. He was almost positive Coleridge was tracking him visually when he got up to ostensibly stretch and pace. While he appeared still, there was a lot going on with this guy.

"So ... do you want to start?" Andersen asked.

Coleridge smiled. "Where would you like me to begin?" The voice was smooth. It had a low timbre.

Well-practiced. Actually, Andersen thought, *his voice does have a therapeutic quality to it.*

"From the beginning," Andersen answered.

"The beginning? Do you mean early in my life or as it related to me being on Summer Street?" Coleridge asked.

"How about a couple of weeks prior to the event if it's related?" Andersen clarified.

Coleridge sat quietly for a long moment. It was evident he was trying to figure out where to begin.

As if he had just seen a starting point, he smiled slightly and asked, "What time is it?"

"8:30 a.m. Does that affect your statement?" Andersen responded.

"No. I was wondering where my assistant was," Coleridge answered.

Andersen waited and then finally asked, "So do you want to tell me how you got to a basement where a shoot-out happened?"

Coleridge seemed to ponder and then continued, "I would really have to go back three and a half years. I can highlight, but it will make more sense to go back that far to give you the whole story."

All right. We'll do it your way, for now.

Andersen's approach to interviewing was to let things run their course at first to give the impression the interviewee was in control and then change approaches halfway through. It was time to start.

"All right. We've got time. Were you advised of your rights? I'm sorry. You waived your rights, I see," Andersen continued.

"Yes. I have not been charged with anything, but I doubt what I say will have an effect on anything."

Andersen noticed that Coleridge actually appeared dark for a moment as if he were remembering something painful.

What was that about?

"No charges yet, so why don't you start talking?" Andersen prodded. Coleridge came back from his dark place, smiled and seemed more composed than he had been just a moment ago.

"Okay. If you got some coffee and arnica, I will tell you everything. And can someone let me know when my medication gets here?"

Andersen got back up and opened the door to call for one of his men and see what they could produce for Coleridge. Andersen knew the coffee was the easy part. Fortunately, his witness had not qualified his coffee as "good" coffee. The arnica was not going to happen. Andersen knew it was an herbal ointment for bruising, but it typically needed to be put on as soon as possible. After a few minutes, coffee was brought to the interrogation room, and Coleridge agreed to a cold compress in lieu of the ointment. Once settled, Andersen returned to his seat.

"So ... how does it all start?" Andersen asked.

"I wasn't always Sam Coleridge," Coleridge began.

"I guessed that. Coleridge was a poet."

And you had to show him that you knew who Coleridge was for what reason? Think Andersen!

Why Andersen had to show Coleridge he knew his poets was unclear to him. Maybe he was feeling the control of the interview slipping away from him.

"Yes. What my name was before is really not important."

Andersen cut him off. So it wasn't halfway through the interview process, but he was getting annoyed.

"No games, Coleridge. There are three dead bodies and two injured federal agents in my town, and I need answers — direct and clear ones." Andersen was now angry. His "witness" was unreal — too calm, too collected.

After a moment of contemplation, Coleridge finally responded, "My name was David Caulfield. I was born December 5, 1967."

Andersen wrote the name and date down. He knew others in the adjoining observation room would be running the information along with his prints as he and Coleridge continued the interview. There was always a technician behind the mirror monitoring the recording equipment in high-profile cases. Hopefully, there would be another investigator in the observation room as well, looking for nonverbal cues. In light of the insanity happening in and around his town, Andersen was pretty sure no police personnel would be spared. Maybe an intern or a cadet would be observing.

"So Mr. Caulfield—"

"No. My name is Sam Coleridge. I am cooperating, so please show me the courtesy of using my name." At that point, Coleridge was silent and as still as a statue.

Andersen contemplated the request for a moment. The more he could get this guy to talk, the better. Sometimes you had to give people in the hot seat some control to run their mouth. He went back to the original plan.

"All right, Mr. Coleridge. How did you end up at Summer Street?"

"Sam is fine. Four and a half years ago, I never would have been there or here."

"All right," Andersen conceded. "Four plus years ago, what happened?"

"It's funny," Coleridge began, "how a former life can be drawn up so quickly from the past and yet your present life is so purposeful."

Oh boy, Andersen thought.

He hated the "philosophical" witnesses. Eventually, Andersen settled with pen and pad in hand, while multiple silent recording devices were meticulously collecting data.

At least the story might be interesting, Andersen thought.

Something to talk to Laura about. She loves this kind of stuff.

She did love the stories, but Andersen really hated it when witnesses and criminals took their time with explaining things.

Coleridge began with the basics — the details of life that are demographic though telling when you compare the past and present.

Chapter 4

"Hoc est verum et nihili nisi verum"
"This is the truth and nothing but the truth"

Present Day – May 2

"IT WAS A LONG time ago," Coleridge began.

Andersen noticed a perceptual shift in Coleridge's expression. He knew enough about interviewing to know that watching for behaviors could impart critical communication, which could provide insight into the person. Up till now, Coleridge was calm and collected. No emotion displayed. That told him that either Coleridge had been in stressful situations before or he was a great actor. But right now, Coleridge's shoulders dropped a bit. He crossed his legs in the other direction, and he stroked his chin.

That was different, he thought.

Andersen refocused on Coleridge's story as he made note of the body language.

"I was married to a beautiful wife and had three teenage children. We had married after her divorce. She was a medical doctor with three children from a prior marriage.

I married late. Loved work more than anything until I met her. We lived in the suburbs, where I was fortunate to be a stepfather to her children. Their biological father made it very easy for me to be the 'nice one' since he wanted nothing to do with them. I had always thought it was because he was an alcoholic, but after meeting him just once, it was clear he had mental health issues in addition to the alcohol problems. Once the kids were in college, we moved closer to the city in a newer development that would be big money someday."

"Where did you live?" Andersen interrupted.

"North Carolina. Then we moved to Virginia. They weren't my biological children, but I loved them. I miss them ..." Coleridge's voice trailed off, but then he regrouped and continued.

OK. *There are emotions there. So that means he has some pretty good control in hiding them...unless they're real strong emotions.*

"I was a therapist then too, but I also taught psychopathology and personality disorders to graduate students and doctorate fellows. My specific focus was on recovering memories and treatment of patients who suffered from post-traumatic stress disorder. Brain development and stressful events could be treated in the most severely traumatized clients, but only if the patients really wanted to. That was key. The brain knows how to protect itself when it's endangered, and it will not just stop 'remembering' something on a whim. It takes time and trust. It also takes physical exercise, meditation, and mental gymnastics, such as learning a new skill. All of this has to be done at the same time you are working on your issues. Most importantly, the client has to be patient."

"I worked with other professors at the university in this approach and would spend some time working at the veterans' hospital's inpatient and outpatient departments with traumatized patients there. I had a small but intense private practice. Because this clinical approach was considered 'eclectic,' even though each aspect of the treatment had its own body of research supporting it, as a clinical practice, it was not covered by the patients' insurance. My practice was a 'cash only' venture that became pretty successful both in reputation and in profits. My clientele were mostly victims of abuse and violence. Sometimes there were high-ranking military and ex-military personnel. I would see them privately. Periodically, there were law enforcement, and most infrequently, I would consult with federal agencies either in an employee assistance vein or criminal case."

"Federal agencies?" Andersen asked.

"Yes," Coleridge answered coolly. "Typically, the FBI and Pentagon staff ... typically in the foreign analyst branch," Coleridge elaborated.

Well, there's the Quantico connection and possibly an organized crime connection too, Andersen thought.

"Go on," Andersen coached.

Coleridge's voice had taken on a more lightened tone, less therapeutic and more storytelling.

"All was going quite well then. The twins were in college. The oldest was in graduate school. My wife was really enjoying her own private practice and being at work full-time.

As for me, I was like all guys in their professional prime — overweight, liked my alcohol —and I was depressed more than less. But I sure made money. I was not living my own treatment strategies for recovery, I guess. But then there was no trauma at that point for me to recover from."

"Trauma?" So you had some kind of trauma that you needed to recover from? Is that where the scars came from? Is that the real beginning of this shit?

There was that darkness he noticed again as Coleridge's story slowed to a stop. Andersen thought he was going to have to prod Coleridge along until he picked the story back up again on his own.

"That was all going to change the day my most complex client appeared at my practice. He was also the most responsive client I had ever worked with. Truly, I was lucky in a sense to see how a treatment strategy on the right person at the right time could change someone. Lucky—"

From Andersen's perspective, Coleridge did not look or act "lucky."

"Does your client have a name?"

"Former client," Coleridge corrected. "His name is Mr. Alexander Burns." Coleridge stopped at the name. He was still but very present.

"So who is Mr. Burns?" Andersen asked. Andersen noticed that Coleridge started with an exhale. Coleridge then smiled, but it was not a happy one, Andersen thought. Coleridge's body language and facial expression seemed to relax and tense up at the same time.

I'm wondering what he's thinking, Andersen thought.

While he was sitting with Lt. Andersen, David's head continued aching right where Burns had hit him earlier that morning. Even though the officers that found him were nice enough, it must have seemed strange to them to find him at the site of such a violent crime. David recalled that the noise, yelling, and shots in the house were horrible to hear when he was in the basement. But it was not as bad as the silence that suddenly followed after such a commotion.

The police made sure he was seen by the paramedics, and then they brought him to the station. He was happy he was not handcuffed. David had no idea what being handcuffed would be like. He had no idea what it was going to be like playing a character while being interrogated by the police. All of this was new to him.

How do Samantha and Alex do this? How do they just pretend to be someone else when they know they're guilty? Do they believe the lie? Is it a lie if you believe?

David shifted in his seat as he refocused on what the lieutenant was saying.

Well, this sure is stressful, David thought to himself. Before he met Burns, he was nothing short of a law-abiding citizen. So honest that his college roommates could not believe it when he called the cable company to report getting free movies. David smiled fondly at that thought. Still though, he was sitting in front of a police officer whom he was deceiving while giving his friends time to create a whole lot of chaos.

If they hadn't taken everything away from me, if they hadn't threatened Emma, maybe I couldn't do this now, David thought.

Sitting in front of the lieutenant in one of the interrogation rooms he had researched was still unnerving to David, even though he had rehearsed this scene several times. Burns constantly coached him.

"Be yourself," Burns would say. "Tell him the truth. It's easier when you tell the truth. Just don't give him the answers he wants," Burns would constantly urge.

David struggled with what Burns said.

For the moment, David was doing his best to seem at ease, but it was difficult. It was not a question of if he could do it. It was an issue of how.

How can I do this? How can I convincingly pretend to be someone named Sam Coleridge? How can I keep this charade going for a couple of hours? he thought.

David knew intuitively that he was the best person to keep the police occupied as the others moved ahead with their plans. Originally, Samantha had been chosen to play the role of victim, but Burns had reconsidered. David had to agree that Burns's analysis was right: While Samantha could stonewall anyone, she could play more roles than anyone outside of the police station. That was her strength — *many things to many people.*

His own skills were interviewing and dealing with people and stressful situations. He also had to agree that he was good at dramatic monologues. So it was with great irony that Burns recommended David initially use the alias of Samuel Coleridge from *The Rime of the Ancient Mariner.* David believed that Burns's choice of name and that specific poem had more to do with himself at an unconscious level. David was sure that Burns's choice was guided by the story's symbolism of burdens, sins, revelations, truth, metamorphosis, and redemption. *That's definitely more Burns than me,* David thought. Still, David followed Burns's advice and took the alias.

Suddenly, David was drawn back into the present with Andersen's next volley of questions.

"Does your client have a name?" he asked.

"Former client," David corrected. It was true — Burns was no longer a client. "His name is Mr. Alexander Burns," David answered.

Then David finally understood Burns's advice about being interrogated. Always give as much of the truth as possible without giving the mission away.

I see now. Tell him the truth.

"So who is Mr. Burns?" Andersen asked.

David exhaled as he realized his best course of action. It would be difficult because he would have to reopen old wounds, wounds that he was still trying to heal, but he had to for Becky and Emma, for Samantha and Burns. David thought back to the day it all started with a phone call a lifetime ago.

Four Years, Six Months Earlier

Jenny was in the shower when David Caulfield, Sam Coleridge's predecessor, took the call at home. Looking at the caller ID, he could easily tell it was work.

What the hell does she want so early in the morning?

"Jesus, Michele! Can't a guy sleep late once in a while?" David snorted out.

"I'm sorry. Did I catch you in your final lap in the pool or on your way to get a donut or something?" Michele quickly shot back.

And I actually pay you for this lip? Good thing you have skills.

"What is it?" David grumbled, still carving sleep out of his eyes.

"You got a new client from the veterans' hospital. No major physical issues per se ... except for the fact he has internal head injuries and burns on his hands and arms ... but seems to have a great deal more mental health issues, primarily paranoia, PTSD, and memory loss. The request comes from the director herself, and the package of information was put together completely by two treating medical doctors and a team from Germany. He came stateside eight weeks ago. Looks like they want you to see him on an outpatient basis as a private patient. Now here is the real strange part—"

David interrupted, "Huh? Something stranger than actually having medical information on a patient from the hospital with no obvious medical issues keeping him there, and the director of the hospital being personally involved? Something stranger than all that?"

Michele went on, "Stranger yet. The director of the hospital accepted your pay fee without hesitation. And they want him in the intensive program for three sessions minimum per week for, and I quote, 'as long as it takes!'"

David moved slowly from the bedroom to the kitchen to get a cup of coffee that had been automatically set to brew the night before. His thoughts had drifted to wondering if there was cream for his coffee when he stopped dead in his tracks at "for as long as it takes." David was quiet. The refrigerator remained open.

That is strange. Anything too good to be true is often something that will bite you on the ass later, he thought.

"All right. What's the client's name?" David asked.

"Don't be too excited. I do forget we usually get clients like this with a potential of 156 sessions guaranteed to be paid within fourteen days of billing with minimal paperwork."

Vintage sarcasm, David thought.

He knew that Michele had to wonder why her boss was not more enthusiastic.

"Okay, okay. I am happy. Now what's his name?"

"Burns. Alexander J. Born January 9, 1984. I booked a two-hour diagnostic evaluation starting at 9:00 a.m. today, so your day starts a little bit earlier than usual," Michele responded.

"Thank you," David said as he pushed the power button on the remote phone and stuck the phone in his bathrobe pocket. Usually, David was angry when his day was changed suddenly. He liked things planned and prepared for. Not this time though. It wasn't because of the money. In fact, the payment schedule made him more nervous of what the hospital might be dumping on his doorstep. He had gotten accustomed to having a bit more time in the morning with his wife now that the kids were all away at college.

"So be it," he said to himself. Jenny was getting out of the shower as he located both cream and sugar and poured themselves coffee to start the day.

"Hello, luvy," Jenny proclaimed. Jennifer was her full name, but she fit "Jenny" better. She was always so happy in the morning that it could be annoying. It was nonetheless contagious, and he could not help but move from grumpiness to happiness whenever she was around. She always made him happy.

"Looks like I am going in earlier than expected." David was blowing on his coffee as Jenny made the final touches on her own cup. She had this adorable approach to pouring the cream and sugar in the cup first and then pouring the coffee in after to thoroughly stir everything.

"Well, that's good. Get in early and come home early. Don't forget you have an appointment with the specialist I set up for you regarding your elevated sugar," Jenny reminded him.

"Yeah."

Elevated sugar. Elevated weight. Elevated cholesterol. Elevated blood pressure. Yeah.

"Doctor, heal thyself," David quoted.

"Please, David. Enough of the drama. You're in your mid-forties, and you work too many hours, don't move at work, and don't exercise. You're not exactly alone in that category."

"Yeah, but I should know better. Part of my prescribed treatment is to have people make time for rigorous exercise to make a breakthrough in treatment. Kind of hard to make that point when you're forty pounds overweight and soft in the wrong places. I don't think I can keep saying, 'I'm big-boned.'"

David continued his complaint, "I really need to make some serious changes."

Jenny listened. She always listened. "Don't worry," she said and touched his cheek.

"We'll do it together."

You know. You always manage to say the right thing at the right time.

David smiled. She always made him feel better.

"Now remember that Bob and Carol plan on seeing us tonight," Jenny added.

Oh, just great! Just perfect!

That just topped off the day of surprises.

"Oh. Now I know it's Thursday … let me guess. The third Thursday of the month?" David bemoaned.

He loved his best friend, Bob; he had known Bob all his life. But after years with his wife, it became clear that Carol was a little controlling and liked to schedule the couple's outings every third Thursday at some new place. Actually, "controlling" was the wrong word. Still, Carol had been the best thing for his lifelong friend. And she kept their marriage new and fresh with all the planning she did.

Maybe I'm a bit resentful of her abilities to take charge and make changes. And I've never seen him happier, David pondered.

"Be nice. She is taking an interest in keeping our friendship alive," Jenny added.

Yeah. You're right.

"Well … time for a shower and to start the day smelling good!" David feigned cheerily. "That's the spirit," Jenny said as she left the kitchen.

When he entered the bathroom, David caught the shape of his stomach in the mirror.

"That's right, fatso. Wash yourself up," he chided himself.

It didn't take long to get showered, shaved, and dressed once he had his coffee. The drive to work was a short thirty minutes because much of the morning traffic had subsided. He wanted to stop off and get a breakfast sandwich, but coming into work with fast food would only prove Michele's point.

He was not about to help her prove any point at his expense. David did like his office very much. It was right on Main Street with a bank of windows overlooking an upper-middle-class neighborhood of artists, small stores, and a surprising number of small restaurants and upscale beauty salons. He liked it much more than his house. His office was in a professional building in the heart of all the activity, people, and energy. His home, nice as it was, was in a private cul-de-sac that was more secluded than he would have liked. The three other homes in his immediate neighborhood were in varying degrees of renovation because the once-working-class and family neighborhood was now changing into new money and really young families. His office, on the other hand, was wonderfully classic.

Michele was manning the front office, and her niece was arranging new and old clients in the schedule to accommodate the new client. But it was not without strings.

"The new client, Mr. Burns, is being brought here by an attending nurse," Michele began.

"Now that is strange. But then the whole case is strange," David replied. "Who or what is bringing him?"

"A nurse, a Ms. Littleton. I think you know her," she quizzed.

Well, that name brings back both memories and ambivalent feelings.

David was unloading his papers and briefcase on his desk as he dug up his memories about Ms. Samantha Littleton.

"Yeah, I know her. Maybe too much about her," David said, more to himself than to Michele.

David saw that Michele let the cryptic statement go.

"Well, it looks like Ms. Littleton will be bringing the file with her and is supposed to be in the sessions with you both in case 'nationally sensitive materials' emerge."

"Yeah ... like that's going to happen," David muttered.

She knows that's not going to happen. Why would she even think that? What's wrong with this picture?

"She wants to meet with you first, so she will be here a half hour in advance of his arrival," Michele continued, and as if on cue, the front office door to the foyer chimed, indicating a new arrival. Like so many times in the past, David had a pretty good idea who it was. As a therapist, David continued to try to analyze why he still had mixed feelings about his former student.

"Good morning, Ms. Littleton," David said while he turned to face her.

She had not changed much since David had last seen her. Ms. Samantha Littleton was an athletic woman in her early thirties. She kept her bleached-blond hair in a functional hairband, and her complexion was clear with pale, smooth skin. While she would be considered pretty, she appeared altogether "forgettable" or so average she seemed to just fade into the background. Other than her unique history, which was known only to him, her own "private clients," and two other mutual associates, she did have one interesting physical characteristic — the iris of each eye was different. She had one blue eye and one light brown eye. That explained why she constantly wore sunglasses outside and tinted glasses inside. Eliminating her unique eyes allowed her to blend in the background more effectively.

"Good morning, Dr. Caulfield. It is nice to see you again."

David smiled but was not sure about her response. To say she was "complicated" and difficult to read was an understatement.

Is she being sarcastic or genuine?

He went with the latter as he ushered her into his office. He really did like his office, he reminded himself. It cost more money, but he had had his office doorjamb customized so that he could accommodate two doors in the same door casing — each closed, giving a feeling of real privacy and security. The heavy blinds on the windows helped with privacy as well. While his furniture would be considered "classic" for psychiatry, he liked the post-World War II decor as a matter of preference.

Ms. Littleton sat in the chair just opposite his desk. While his desk was somewhat littered with papers and files, he could find everything he needed, much to Michele's chagrin. She was into color-coding everything. David removed his briefcase from the desk so that he could sit down and actually see Samantha as he talked to her.

"It's been a long time, Dr. Caulfield," she began, removing her sunglasses in an uncharacteristic gesture of openness.

"Just about six years. I did hear about you graduating nursing school and passing your licensing for your RN."

"Your written recommendation did get my first job at the veterans' hospital in addition to college. I always had the sense you were watching me but in a good way. Do you take care of all your former supervisees this way?"

David could tell it was hard for Samantha not to present herself as seductive. It came naturally to her. He had never acted on her seductiveness and had always been a gentlemen, almost a father figure. He suspected he was a significantly different one than she was used to.

"Actually, yes," David said honestly. "There have been a number of children and adolescents, former clients, I have assisted in getting into college and graduate school. Periodically, I have assisted in helping them get their first jobs, but only if I thought I had something to really offer and as long as it did not violate boundaries. You were the first adult I had assisted in such a way. I think it turned out all right ... though your original choice of profession has confounded me at times," David offered.

"Well ... I did have a history of providing a service to men for a few years prior. When I chose psychiatric nursing, I thought it would be along the same vein, helping those in need for medical and health issues and not just sexual," she explained.

OK...more information than I needed. Now you're here because...

"Though truth be told, Dr. Caulfield, I have increased my part-time job. I make three times as much in half the time compared to a full shift of nursing. A noble profession, nursing, but it just doesn't pay," Samantha concluded.

David had always thought that. She was one of his successes that he never talked about. He did appreciate that she always paid for supervision in cash and was remarkably intelligent, but her options in life were limited when she was younger. Her getting sick with cancer and subsequent depression allowed her an opportunity to stop and look at what she really wanted to do with her life. He always recalled that her plan was to "have the power to redefine myself." He knew a lot of smart people in his life, but her insight into her own future was startling in brilliance for one so young. She did use treatment, and not only did she go back to school and get her degrees, but she also changed her profession ... though not entirely. David was aware that she always kept a number of well-screened clients that paid very well. She was very courageous to make such substantial changes in her life; however, she was also paranoid, and she liked having a backup plan. Prostitution allowed for extra money, flexible times, untaxed and unreported cash flow and allowed for an alter ego. Sam Littleton's alter ego was "Danielle Spencer."

"So I take it that Ms. Danielle Spencer is alive and well," David attempted to confirm.

"Yes. She not only keeps my finances in order and allows me the opportunity to volunteer at the VA, but it satisfies my sex drive," was Samantha's short answer.

Yup...way too much data. I wonder if she would do this if I was a woman? An elderly woman.

"Wait a minute. You are paid at the hospital, aren't you?" David asked. He was positive the nursing positions paid pretty well at the federal government level.

"Well, $42.20 per hour that is taxed compared to my 'reduced' hourly rate of $250.00 cleared. Well, I think you can see why I might consider my nursing gig a volunteer position."

Wow! That is quite a range? Still, I'd rather you stay with nursing. Logical, no. The numbers and your reasoning are sound.

He decided to take another approach.

"Well, Ms. Littleton, you look well," he said.

"I am still getting rid of the last twenty pounds from the prednisone, but I am nearly back to my old self," Samantha concluded. "So what brings you here, and what makes you think I am going to conduct treatment with you in the room?" David pressed. He knew that Samantha knew that he would not allow her in his sessions.

"I know you weren't going to let me actually sit in on the session. In fact, I just wanted to give you a major heads-up about your new client. When he says he doesn't remember what happened, I do believe him. But he knows that he knows a lot more than he recalls, and he is desperate to find out. And so is a bunch of people in the intelligence community and the Department of Defense. I was the triage nurse on call and was the first to meet with him. He had all the classic signs of trauma-induced amnesia ... not that I am trying to do your job," she said and smiled.

You know, you have a great smile. You've always had a great smile. Rare but radiant.

"I'm glad you left me something to do," David replied. "Please go on," he urged.

Samantha handed him a full folder of information on his soon-to-arrive client. There was an extraordinary wealth of data, though it was unusual to have this level of information before a session. Samantha knew that too. She continued with her report in the way she always had when she had interned for psychiatry many years ago.

"I have never seen such a wealth of data and history available, and yet it doesn't seem to fit him. They have his life all laid out in a neat narrative. More importantly, he does not seem to 'fit' his own history. He seems more depressed than extroverted, more reflective than a party animal. And when you dig a little bit into his past, it doesn't add up."

Samantha became quiet and then continued, "A private client of mine who graduated the same year from the same college did not know him. It was a small enough school. A picture of Burns from his yearbook compared to my client's did not match up."

"Did you tell anyone about this?" David had to ask.

"No," she replied.

Now why am I not surprised at that?

David knew why — her survival instincts were always online. She could read people and situations better than anyone he knew. If she thought something was wrong with this client's past, he would bet she was right.

"Well, I gotta run," Samantha said as she began to stand up. "I don't want him to see me."

David was genuinely surprised.

"I thought you were going to bring him back and forth and were planning to sit in on the sessions?"

Thinking more, David went on, "Wait a minute. Why are you so interested in this client's well-being?"

David waited as Samantha slowly answered, "He is different from the other veterans. He had a head injury, and usually, we bring them off of the majority of the medication they are on once they are stabilized to get a baseline. They didn't do that in his case. In fact, I swear they kept him on a whole bunch of medication to keep him sedated."

"Hmm. So why is he now coming to see me?" David pushed.

"I think because the director of the hospital wanted

some movement in his case and you were recommended," Samantha said.

David didn't want to know how that happened, and Samantha did not elaborate any more. He was often confused by her behaviors and actions. She was a survivor but seemed to be driven toward fair play. Someone was victimizing someone, and she was going to stop it. This ran against "self-preservation at all cost" which Samantha seemed to constantly project. He knew she had been abandoned early in life. After years of knowing her, he eventually discovered that she had been in a number of foster care homes until she finally got to the one that had the only person he was sure she loved, her sister, Becky.

This sister had to be someone special, he thought often. *Maybe a caretaker? Maybe an actual mother? Protector? She has to be someone Samantha intuitively trusts—someone real, genuine, and not fake.*

He knew her family was financially strapped, so prostitution for her was not only a way to make money but also a way to connect — at least physically. That made David sad to think about. He envisioned a young, vulnerable woman trying to make connections that had time limits and prices attached.

Always conditions. Poor thing. That just ain't right.

David remembered having lunch with a female colleague and Samantha joining them ever so reluctantly. While he was struck by her social awkwardness, it was apparent from the conversation which his colleague expertly extracted from her that having lunch, dating, going to a movie, or having a swim — all these normal, everyday events — were simply alien to Samantha. *She just didn't have a normal or easy life*, he had often thought after that revealing lunch. And still, after all that, she wanted to help some guy she didn't even know.

Remarkable, David thought. David pulled himself back to the conversation.

She's not even going to say hi to him? he wondered.

"So you're not even going to stop by to see him?" he asked further.

Samantha smiled again. She folded her arms, and her eyes seemed to get unevenly dark.

"Dr. Caulfield, they removed me from the case two months ago. I asked too many clarifying questions. I was then transferred to another ward. I 'collected' this file as it was the one they brought in from the field, and I am betting the one that the attendant will be bringing you will be somehow 'different.'"

She continued, "Over the last month, I have spent more time at my other work address in the evening and seemingly less time at my home and the hospital. I go to work, business as usual, but I am certain something is wrong there and I am being watched." Samantha handed him a simple cell phone. "This cell phone has my work cell number and information on a meeting place where I might interview new clients. Don't call me at my day job. Right now, I am going downstairs to my massage appointment a girlfriend has set up for me. I am sure they called to confirm I am there now. Good luck."

And as simply as that, she walked out, sunglasses on, backpack slung over her shoulder, looking like so many older students or young moms in the neighborhood. She was dressed to fade into the background.

It was rare for David to be speechless. This was one of those times.

Well, now, that's very strange. No more to talk about, so you just leave. How can you just do that?

It was surreal and completely foreign to him. While he had seen some strange stuff both within the bounds of the law and beyond it, this situation seemed very uncomfortable and covert.

She couldn't just give me a cell phone number but a whole cell phone instead, David thought.

"Michele? When my new client gets here, make no mention of Ms. Littleton's arrival please."

Without hesitation or question, Michele confirmed she would not recall.

He was still pondering all that had transpired when the door chimed again. This time, it was a young brunette woman with dark features. It was evident that this nurse was not fully comfortable in her white uniform. She also held a large case file in her hand and David's new client in tow. She was not pretty but also not homely either. "Strong" or "athletic" would be the best words to describe her.

Maybe she's of Mediterranean descent. She's something.

Her mannerisms seemed all business to him and there was something about her eyes David didn't like.

Dark. Eyes with little to no reflection from them. Not exactly nurturing for a nurse, David thought.

"Good morning, Dr. Caulfield. I am from the veterans'

hospital. Here is Mr. Burns's medical file." The room seemed too quiet, and she became uncomfortable suddenly. David did not help with the uncomfortable situation as he started reading the new file, simply standing quietly in front of her and his new patient.

"So should I come in with Mr. Burns, or should I remain outside?"

David came out of his reading trance and returned to the here and now.

"Yes. I mean, yes, you can remain in the waiting room. Michele will let you know the best places to get coffee. I typically do my work with my clients without observers. If that is a problem, I am sure there are a number of qualified clinicians that could see Mr. Burns."

The attendant was surprised but not as much as Michele. David knew there was a lot of money that was being pushed out the door. His new client remained quiet, but David noticed he was smiling just a little bit. The attendant recovered quickly, "No worries. I will be able to read." Turning, David could see Michele's eyes close with obvious relief that he did not lose their economic windfall. Once settled, the attendant sat down. Quiet, broad-shouldered, and somehow present and invisible at the same time, Mr. Burns walked slightly ahead of David and took the seat closest to the door.

Still looking at the new file, he could easily see the similarities and the differences.

Two files? One fabricated and the other with more details? Why? Who would do this and why? Is there a real conspiracy going on here?

"A double door? Is it for noise or security or both?" Mr. Burns asked.

David became self-conscious of the first file he had been reading, the one that was still open on his desk. He took a moment to walk over and close the other file, and he then opened the "newer file" in front of him now.

"More for sound than security. Much more for reassurance than anything," David said in a distracted manner. He was reading the file again, and while it seemed the same, there were some differences: grammar, sentence structure, and slight changes in the dates, especially in regards to the college years.

It became apparent that David was being rude, simply focusing on the case file, when Burns asked, "So how do we get started?"

As if awakened out of a sleep, David made a decision. He put the file down, took out a pad of paper, walked around to a chair beside his new client, and said, "Okay. Let's start."

Burns asked, "Where would you like me to begin?" David responded, "At the beginning. What do you first remember?"

Chapter 5

"Omnia mutantur nos et mutamur in illis"
"All things change, and we change with them"

Four Years, Six Months Earlier

BURNS WAS SITTING IN an oversized, leather chair. Unlike most of his clients, he did not settle back into the chair. Rather, he sat more on the chair's edge, his feet firmly set on the floor, his hands folded on his lap. Burns wasn't wringing his hands per se, but they were nervously moving.

"I remember waking up in a hospital about six months ago. I remember my hands were bandaged up and my eyes were covered. I remember the smell of smoke from electrical circuits. I also remember having difficulty breathing as if I was underwater."

David noticed scarring of second-degree burns on his hands and slight scarring near his eyes and nose. There was new hair growth on the back and right side of Burns's scalp, indicating either recent head or brain surgery. Burns's cadence and voice were low, but his eyes were everywhere in the room. The eye movement was not that of someone nervous but seemed more purposeful. David asked, "What are you looking for?"

Burns stopped and focused on David.

"I am just checking out your office. I do that as a matter of course. I am not looking for anything in particular."

OK. I may not be security-savvy but I know when a trapped animal is looking for an escape route.

David knew otherwise. "No, you are looking around for something. Either a vantage point or an escape or both."

Burns smiled. "Sorry. A job hazard ... so I am told."

David noted the smile. It wasn't a nervous smile or a fake one. The smile seemed genuine. The corner of the eyes narrowed, and the nose seemed to flare slightly.

"And what were you told your job was or is?" David queried.

"They tell me intelligence." Then Burns shifted quickly. "So, Dr. Caulfield, is everything I say in here confidential and private, or will you be reporting to the hospital?"

Now that is an excellent question. So he does have both genuine emotions such as humor, and he smiles in addition to displaying intelligence. There's a lot to work with here, if we work together.

David took a moment to respond. The usual response was to review confidentiality and privacy and duty to warn in case of danger to self or others. David decided to take a different approach.

"Mr. Burns, I would like to say yes, but I'm simply not sure that is the case. With the director of the veterans' hospital watching, an extensive file immediately available, and a whole bunch of other interested parties that seem to have a vested interest in your memory, I really can't say if anything you say or do will remain private. I'm guessing anything I write down or hear will be documented, seized, and questioned at some point. So if there is anything you don't want others to know, you may not want to tell me."

Well. That's pretty honest. A long answer to a short question – everything you say will become public, so watch what you say.

Burns seemed very surprised at either the bluntness or honesty or the balance of both.

"Well, you are the first person I have heard tell me something I never thought I would ever hear from a therapist — bluntness and candor."

It was David's turn to smile before he went on.

"I hope that you will tell me things that will assist in my diagnosis and treatment strategies to help free up your memory. I am guessing your brain is blocking memories either as a way of self-protection or a trauma interfering with recall. I have no vested interest in the content of the data. Just being able to assist you in accessing your memory is my plan. So if the local intelligence community decides that they want to have any data obtained from our sessions, I will ask them to produce a court order and hand it over rapidly. If you plan to harm yourself or someone, I will tell everyone on the planet to ensure your safety. I just want to be clear with the plan and limitations of treatment," David concluded.

Burns's look was thoughtful and still.

"I like your honesty and approach. If there is something I don't want anyone to know, I won't tell you. So can I ask you a question, and will you be honest with me?"

David answered cautiously, "If it pertains to me and my family personally, no, but otherwise, yes."

"Did someone else bring you into the loop before I got here? Have you seen my file already?"

How could he have possibly known that? Did he see Samantha outside? Wait a minute...Easy, David thought.

As a therapist, he was pretty good. As a covert spy, he was awful. The first file on David's desk looked nearly identical on the outside to the one he now held. Because all the other files were smaller and color-coded, anyone with good eyesight could see the similarities in these two separate files.

David took a moment to consider if lying was an option or if telling the truth to his client was appropriate. He decided to tell the truth for a couple of reasons: he knew he was terrible at lying. It took too long for him to answer the question, and he was positive that Burns could sniff out bullshit from miles away. David settled even further in his own chair as he folded his own hands and crossed his legs at the ankles.

David answered slowly: "Yes. I have an entire file that was 'collected' seemingly unofficially and brought to me by outside channels. There was concern that either data would be altered or history rewritten."

David felt bad. He had thrown Samantha under the bus in order to obtain the trust of a client.

"Were there any obvious changes?" Burns went on.

"It looks like the college time was altered, but I didn't get further than that. I had just got it moments before you arrived."

Burns smiled. "I knew I liked that nurse," he said more to himself.

David was somewhat relieved. Clearly, his new client had his own knack for figuring people out. It looked as if Burns saw Samantha more as an ally than as an enemy. David remained silent as Burns scanned the file he had just handed him.

Burns closed the file and then seemed to make his own decision.

"Okay. I will tell you all I remember. I will ask you to compare both files and just let me know the inconsistencies. There is something big going on in my head, and I really can't figure out what it is. I really do need your help."

David noticed that Burns became distracted, as if something was just in reach of his memory but was fading fast.

"What are you thinking?" David asked.

It looked to David as if he was looking for the right words. David noticed Burns was lightly scratching his scar tissue on his scalp. Still, Burns seemed determined to answer David's question.

"I have this strange feeling that I used to be a part of something important. I think I used to be a soldier. I think I used to help people or at least I protected people from the bad guys. It's something important, Dr. Caulfield. I just can't get it on my own."

Damn it, David thought.

His client was genuinely asking for help. Part of David was hoping that Burns would be resistant to treatment. If Burns was not open to treatment, David would be able to convince himself that he had tried and the patient was not interested in his help. But because Burns was open to help and had actually asked for assistance, his natural desire to help had been triggered. David knew that when he invested in a patient, he would take treatment as far as possible, no matter where it went.

Maybe that's why Samantha had him sent to me, he thought.

"So how did it go?" David heard from outside his line of sight. It wasn't a new client but a disembodied voice that seemed to be in the office but invisible.

Wait a minute...where am I...

Present Day – May 2

Andersen could plainly see Coleridge was still remembering that fateful day he took Burns on as a patient. Coleridge responded as if he was hearing a disembodied voice of a ghost. While he was blind, Andersen watched as his witness instinctively blinked and rubbed his eyes as if he was in a trance while he was trying to return to the present. It was evident that Coleridge was physically trying to shake off his former self.

Recognition seemed evident on Coleridge's face. His cheekbones and shoulders relaxed. He shifted his legs to cross in the other direction; his hands returned to his lap. Yes, Andersen could see that Coleridge was back with him in the present, being interviewed by an investigator at the North Reading Police Department.

"So how did it go?" Andersen repeated as a way of ensuring that Coleridge was firmly back in the present. Andersen knew that the ice pack was warm by now and Coleridge's bump had to be throbbing less because his statement was moving at a pretty fast clip. Andersen was surprised that his witness had drunk the coffee. Most people had one or two sips and were done. Some criminals confessed to their crimes after they were only allowed warm water and cold coffee from the police station to drink just to end the pain. Andersen smiled for a moment. He had told Dempsey that theory in jest, but Dempsey told others as if it were an actual police tactic. Andersen's train of thought returned as Coleridge started to talk again.

"Burns was quite the specimen, both physically and mentally. It took weeks of first preparing and then actual treatment." Coleridge slipped back into professional mode in explaining the treatment approach.

Coleridge's visage remained still and relaxed, but now he leaned forward and folded his hands on the desk.

"The approach for many of my traumatized, anxious, and to some degree depressed clients is to use their own body to alter brain chemistry so as to decrease the barriers and unlock the guilt, secrets, and problems. In Burns's case, he had the will and healthy physical baseline and time to accomplish this. I generated a workout plan to get his heart rate up, blood pumping, and body moving so that I could get his stress down, the good neurotransmitters going, and get his vegetative signs more stable."

"Vegetative signs more stable?" Andersen asked.

"Yes. In other words, his sleeping and eating patterns were way off, so we needed to get him on a regular eating and sleeping cycle again. This included a decrease in carbohydrates and a significant increase in proteins, fiber, antioxidants, and more 'brain food' like fish and fish oil. I also added meditation three times a day for twenty minutes each day to be said aloud."

Andersen was really curious now.

"Meditation? Why?"

"The neuroscience of the time indicated that when a person meditates for fifteen minutes every day, their frontal lobe rearranges itself. The frontal lobe is known for its abilities to think logically, plan, strategize, and is considered to be the 'computer part' of the brain. The frontal lobe also tends to be pro-social, positive, and overall empathetic. If you alter the frontal lobe with all of these efforts, it becomes the way of regenerating neurons and to have the person think more positively. Maybe even happier. And happy people are less guarded and more open to talking, sharing, and remembering things."

Andersen stopped writing because he was getting lost.

"In English please?" Andersen asked.

Andersen watched as Coleridge's body language shifted.

More animated, relaxed, and confident. I can see this guy is a natural professor, Andersen thought.

Andersen could see that he was trying to put it all into layman's words.

"The part of the brain that makes us thinking creatures not only helps us think but helps keep us civilized. Laws are important. Codes of conduct are important. This is where our moral basis is consolidated. That's why prayer, learning, meditation, and positive thinking make for happier human beings," Coleridge concluded as he adjusted himself in his seat.

Yeah. This guy is a good teacher. So how did a teacher and a therapist end up a thousand miles away from home in a small New England town in a fire fight?

Andersen had to ask, "What did he meditate on? A Hindu prayer or Indian chant?" Andersen asked, getting back to his interview.

At the very least, Coleridge or Caulfield or whoever he is was interesting, Andersen thought to himself.

"That was also a shock. Burns chose Eastern Orthodox prayers from the Orthodox liturgy; he chose morning, midday, and evening prayers each for thirty minutes every time. That's an hour and a half a day when only twenty minutes would have been helpful. In addition, he took the Lord's Prayer and Hail Mary prayer and meditated with those as well in Latin. Latin of all things. So in addition to a remarkably rigorous exercise routine and improved diet and sleeping patterns, he was spending about two hours a day meditating on religious matters that focused on forgiveness, acceptance, and kindness. And he had learned a new skill. He was reading Latin. This entire program took about sixteen weeks. That was strange."

Latin? It took me four plus years to read that dead language.

"Strange?" Andersen asked. This was one of those situations when he thought more would be better.

If only twenty minutes was needed to meditate, wouldn't two hours have to be better?

"Strange? Yes. I expected to see results in four weeks. With the rigorous exercise he was doing, the average patient would start feeling significantly better and their mood greatly improve while their anxiety decreased. This would make them more amenable to therapy. They would start talking more freely and with far fewer barriers. Burns didn't start talking in a less guarded fashion until week eleven. But that was minimal at best. More time passed. And even then, after nearly four months of this level of treatment, his breakthrough seemed more like a surprise."

"How so?" Andersen encouraged.

Coleridge sat back in his chair again as if recounting the very moment it all snowballed. Andersen could see him returning back in time. His shoulders seemed like they were tensing again, and even his voice sounded a bit weaker and hollow to him. Coleridge started to talk again.

"I'm a computer geek. I love tinkering with old desktops. I build computers up from the motherboard to RAM to multiple monitors. Right before one session, I had soldered two components. I had thought that after two hours, the smell would have evaporated from the room. But Burns had picked up the smell. He came in the office as usual, but he seemed to tense up immediately and become focused. He had changed almost before my eyes. His posture seemed to alter. His voice dropped to a lower tone, but it was empty at the same time. His expression seemed measured, and he seemed to almost read my movements. But he seemed distracted at the same time. It was confusing to me. It was more than just his physical changes from his exercising. It was as if all of whatever was on his mind became very clear to him."

Andersen noted the irony of Coleridge's description of Burns changing before his eyes as he was watching his witness go through the same metamorphosis when he thought back to his former life as David Caulfield.

Wow. This is weird. Like you're actually going back in time.

Andersen wondered how Coleridge managed living two

separate lives in one body in two different time periods — the past and the present.

It has to be draining.

Four Years, Six Months Earlier

The lieutenant's question is a good one. Not a simple one but a good one, David thought. He would never forget the moment he saw Burns change. David had seen such transitions before as someone would finally remember a lost memory or make a pivotal connection.

All because burnt wires triggered his memories. That was the trigger, he remembered.

David had since reflected that Burns's complete metamorphosis was the beginning of his own, one from honest therapist to a criminal. Easy to recall. *Painful.*

"Alex? What's wrong?" David asked.

David could easily see that his client's eyes were burning hot though still and focused. Typically, Burns's eyes were soft brown. Not now. Burns shifted from a searching, distracted mode to a suddenly focused presentation.

Then he uttered an odd question: "Do you know what the ace of spades means in the intelligence community?"

"Ace" as in "ace" in a set of cards? That's really out there.

Burns's casual question was out of sync with his body posture, which was now still and vigilant, listening for any possible noise as his eyes remained fixed.

Well, this is very startling. It looks like he's in a dissociative state.

David knew that in these cases, remaining calm was

critical to keep the client grounded in reality by talking. Talking in a calm voice was the first step in bringing his client back to the present.

"No idea. I thought it was a suit in a deck — the highest card for that suit, actually," David answered.

What is going on here? David wondered.

Still transfixed and standing still, Burns went on as if he were sitting in his chair, talking to an uninformed friend.

"Using playing cards for identifying bad guys goes way back in military history. The ace of spades was the card for Saddam Hussein in Iraq, for example. There are still eight of the original fifty-five still at large. Did you know that?"

David sat down as Burns remained standing near the door. In addition to his soothing voice, he thought that if he sat down, maybe Burns would as well, or at the very least, he would appear less of a threat if he was sitting while Burns remained standing.

"What does this mean? Is this the thing you have been trying to remember? Is it coming back?"

Burns remained silent, recounting something.

"You were a soldier? A field agent?" David offered in an attempt to appear relaxed as he sat back in his chair. David could tell that his presentation of relaxation was not very convincing, though he was sure Burns's mind was not in the same room with him.

Definitely a dissociative state. Or maybe he's being flooded with images of the actual trauma? Maybe it's a post-traumatic reaction? That seems more likely based on the level of intensity. Though I would think these thoughts would be overwhelming?

"No. More than that. Logistics primarily. I also had

communication experience but to a lesser degree. I had more field experience than the usual logistical and tactical specialist, but information, recon, and data collection was my specialization." Burns was far away at this point, but not as far as he had been a moment ago. Burns seemed to be returning as he moved two steps toward the back of his own chair and placed his hand on it for support.

Is Burns going to pass out? David wondered. *Is there a secondary issue from the brain trauma that might be affecting his balance or worse?*

"I did many bad things, Dr. Caulfield. It is disheartening to remember all these things I did," Burns trailed off.

Guilt. That's in there too, David thought.

It was easy to see that Burns was remembering very difficult things about himself that had been locked up tightly in his brain, David surmised.

Knowledge causes pain. Maybe your memories were locked up for a very good reason. Problem now is that we can't go back.

Then suddenly, Burns was in the here and now, offering a partial answer as he stared intently at the windows.

"I worked with an extraction team. And my ride was hit, and we crashed. I remember my team continued with the mission without me, but there was so much more. Do you remember your colleague who provided the first record?"

David was caught off guard by a question being fired at him.

"Yes," was the short answer.

"She may know more than she thinks, and they might think she knows more than she does. Either way, it's unsafe for her, and you should let her know that she should not return to work."

Burns quietly turned toward the door and left. It was a full minute before it was clear that Burns had just walked out of treatment. David had had patients storm out of his office. They had wept when they had left. David had even had some patients wheeled out of his office in full restraints to be hospitalized before. But this was a first.

No. Not a first. Samantha had done the same thing months ago. No good-byes. Maybe those two are more similar than different.

"No ... thank you. It was nice working with you," David said to no one.

"Michele!" he finally bellowed.

"My God! What are you yelling about now?" she responded as she came through the same open doors that Burns had just exited.

"Don't buy the leather couches just yet. Our cash cow just left without so much as a 'thank you.'"

Michele frowned. Her eyes narrowed as she peered at him. It was easy for David to figure out what was on her mind.

"And for the record, I did not make him go, either. It's as much a shock for me as it is for you," he added.

"Call the hospital and let them and whoever brought him today know that he just walked out."

David got up and walked to the doors and closed each as Michele exited. He found the cell phone Samantha Littleton had given him in his desk drawer, and now he planned to use it. He dialed the only number in the contacts section and waited.

Samantha's voice came on, though it was very different.

"This is Danielle. How are you, honey?"

It took David a minute to remember Samantha's primary career. The voice was light and seductive. It made you feel like you were the only one she knew.

She does that well, David thought.

"Samantha? It's David. Our mutual client wanted me to let you know that you should not return to work and that your safety is of concern."

The alter ego on the other end of the line suddenly changed to a familiar voice.

"Okay, memorize the number and address and get rid of the phone. It's been nice knowing you, David. Take care."

"Samantha, whatever is going on, you need to be careful. This thing seems a lot bigger than anything I have ever imagined," David pressed.

"Will do," Samantha added briefly. Then the line clicked off.

David closed the cell phone and remained standing for a moment, looking at it.

All right. Now how do you get rid of a cell phone? Throw it in a river or something? David mused.

"I'm not a spy," he finally said, exasperated.

Should I take the SIM card out of the cell and flush it down the toilet before I toss the phone itself? he thought.

Well, I got something done right with originality. SIM card comes out and tossed in one place, and the cell in the river. I think after this I'll stick with therapy and teaching – something I'm good at.

He did plan to drop the cell, but in pieces on his ride

home after he had flushed the SIM in the toilet. For now, the cell was in his coat pocket. Well, at least Burns was his last patient of the day.

Time to go home. I hope he's OK, though, David thought.

It was amazing that no more than ten minutes ago, there were two people in his life, and then they both just walked out, never to be heard from again.

Surreal, David reflected as he collected his briefcase.

Part of finishing work for the day meant that as David would get ready to leave his office, he would reflect on his patients. While Burns and Samantha were the last two people he had to deal with clinically, he started his evening ritual by going over the cases he had had earlier.

With his car right next to his building in an open parking lot, it was a short walk to enjoy the setting sun and a beautiful, warm night. With patients running through his head, David thought he would call Jenny to meet for dinner rather than just head home. Suddenly, his happy thoughts vanished as he saw two men blocking his path. They were in dark suits, and they were pretty set on getting answers.

Well...so much for a beautiful night and meeting anyone for dinner.

One man was very large in height and width. The other agent was smaller but reminded David of an armadillo.

That's an odd way of describing someone who is short but strong, David thought.

"Dr. Caulfield?" the larger man asked.

"Yes," he responded promptly. He still had the cell phone in his coat pocket, the SIM card in place. He was still feeling guilty for not getting rid of the phone.

You know, the whole point about having a plan to destroy something is to actually do it and not carry it around on your person to do later. Clearly you wouldn't make a good career criminal, Caulfield! Yeah...stick to clinical work.

"Dr. Caulfield, we are from the US Department of Defense's Foreign Intelligence Office. We would like a word in regard to your patient Mr. Alexander J. Burns." Their badges and identification confirmed their employer.

"Well, you just missed him by about fifteen minutes," David answered in full knowledge that he had violated patient-doctor confidentiality.

Both men looked at each other and then back at him. They waited for an answer. While the sunglasses obscured much of their facial features, it was also evident they were not happy with this new development.

"Where did he go?" the large man asked as the smaller man immediately stepped away from both of them and began low, curt discussions with his cell phone.

"Honestly, I have no idea. He clearly remembered something of importance. He asked me if I knew what the 'ace of spades' meant, and then he simply left. He gave no clue as to what happened and where he was off to. I did notify the hospital about him seemingly going AWOL," David concluded. David then looked at the larger man. The larger man's face was without expression, and not seeing his eyes made him feel uncomfortable.

The man was not too happy, but it was clear his anger was directed at something else.

Then from behind, David heard the voice of the attendant who had brought Burns to his appointment.

The attendant was a man this time with clear muscular definition.

"Is it true? Did Burns just take off?" the attendant asked.

With his attention now divided, David turned to the new voice, acknowledging the accuracy of the statement. The attendant got on his phone and was now speaking in a quick, urgent tone as he picked up his pace in the direction David had planned to take to get to his car.

Straining to hear what the retreating attendant was saying, he could barely make out one clear sentence:

"...Get your ass over here and have Webber contact Daniels!"

David's attention was brought back to the larger federal agent again who began talking to him.

"We'll be back, Dr. Caulfield. If you hear anything from him, call us as soon as possible. This is a matter of national security. We will have a court order for your records. That means all notes," the federal officer added.

Okay. Maybe I caught a break here. Once I ditch these guys, I'll be able to get rid of the cell phone. I have to get rid of the SIM card first.

"Do you want it now, or will tomorrow be all right?" David asked.

David truly expected an answer such as "a day or two."

"Oh, not right now. It will take about thirty minutes for the court order, and we will need to supervise the process. If you want to get a cup of coffee with my colleague here, that will be fine."

David frowned.

So much for catching a break. So much for dinner with Jenny.

"All right. Why don't we just get this over with and start packing," David answered.

"That's a great idea," the smaller federal officer supported. David gave a quick look back, and the attendant was out of view. He was surprised how rapidly the guy had vanished. He was also surprised the federal agents never even seemed to acknowledge that the attendant had ever been there.

They have to know each other.

As David walked back to his office with two federal agents behind, he was still worried about Alex.

Why is everyone after him? I hope he's okay, David thought.

Chapter 6

"Culpa"
Sin

Four Years, Six Months Earlier

BURNS WATCHED EVERYTHING FROM the shadows of an alley across the street, less than a half-block. He was sure no one else would use the alleyway as a shortcut as the stench from the dumpster was powerful.

Just like old times, he thought as he watched commuters passing back and forth.

Getting out of the building undetected was the easy part. Getting across the street without the federal agents, the attendant, and driver watching the building was not. The federal agents, one larger and the other one more compact, were at two different vantage points across the street, while the driver remained in the car in front of the building. The attendant remained just in sight of the front door on the sidewalk. After five minutes, a fifth person arrived and entered the front passenger side of the car. She was a stocky woman with dark hair and dark features. Her clothes stuck out because of how blatantly they cried out "security detail" with the vintage gray slacks, white blouse, and blue blazer. This was the first attendant that had brought him to see Dr. Caulfield for the first several visits.

Well, you're back again, Burns thought when he confirmed who she was.

Burns had a visceral reaction to her. Her movements were efficient. She seemed to show very little emotion and always seemed to be looking for advantages, physical or otherwise. Burns had seen this kind of woman before.

No. You're not a soldier but a hired killer. Movements economical and efficient, eyes everywhere, walk deliberate...you definitely have a plan.

He didn't miss her and was glad she was gone. Burns knew that seeing her now meant there was serious trouble to come. After ten minutes, there was sudden movement. As the world of bystanders moved all around, they seemed to dissolve as Burns focused on the five people of interest.

The two federal officers moved toward the professional building as David Caulfield exited. As they approached, the person Burns knew as the attendant watched, and the two in the car remained still and observant. There was an exchange between the doctor and the federal officers; the smaller one turned to make a call, while the other watched the doctor.

The attendant came up and clearly asked, "Did Burns really take off?"

Once answered, the attendant turned and talked on the phone to someone, but it was not the people in the car as it was evident that they did not appear to be listening to their cells. The attendant continued talking as he slipped in the back seat of the car. The car remained parked as the two in the front seat listened to their side of the conversation and looked forward.

Now who are you talking to? Webber? I bet it's Webber.

Burns turned back to the federal officers and Caulfield, and then he saw them all return to the building. Burns assessed that they would probably collect all the doctor's notes and grill him for a couple of hours. Burns liked Caulfield; he knew he would cooperate and give them the data, and that would tie them up for hours. The crew in the car was another matter.

They're not military or federal, but they're something — private security and professional. More than babysitters for sure.

Burns focused on the car. The phone conversation was now done, and the three in the car were now talking. The two in front listened as they continued to look forward, while the attendant in the back seat continued to talk to them and issue orders. All seemed matter-of-fact until the female occupant shook her head as if she understood something, exited the car, and hailed down a cab.

I wonder where you're going, Burns thought to himself.

Burns had an overwhelming urge to follow the woman. He knew wherever she was going, something bad was going to happen. Pushing his feeling aside, he tried to focus on his acquired skills and experience as a field agent. Burns shifted his attention back to the car's remaining occupants. The driver handed the attendant something. There was some brief discussion, and then the attendant exited and started walking away from the car and building. The driver then carefully pulled out and was gone. The decision was clear — he would follow the attendant to get information. This guy seemed to be the one issuing orders and seemed to be in charge.

It was a warm and beautiful evening. In any other circumstance, Burns might have appreciated it, but there was a job at hand. His immediate mission was simple: data collection of who was on to him, how much they knew, and most importantly, who was running all the missions. Not just this operation, but all of the operations Burns now recounted. The dates and locations were murky still, but the intelligence, data, objectives, and results were crystal clear.

Are the players still the same? Am I designated for termination? You'd think they would have just done it by now...unless they need something.

For Burns, the files stolen, the targets killed, the countries destabilized, the civilians caught in the crossfire, all of these facts came back to him.

Foley? Deb Foley, he thought at first. Then he remembered a minivan with a mother and child being used as a weapon to stop a terrorist.

Yeah...that was me, too. Anything for the mission, he bitterly thought.

While the images were disturbing, Burns was mostly concerned about something else. He had two opposing feelings that seemed to flood his thinking and entire being.

First, he felt great regret for his past behaviors. Secondly, after he had absorbed all of the images, thoughts, and feelings, he felt a great weight hang from his neck. He felt as if he could barely breathe. These feelings had been so overwhelming that he had thought he was going to fall over in David's office. Burns remembered telling David that he thought he was part of something important, that he thought he used to protect people from terrorists. As the memories returned, he was partially right. He did protect people in the beginning. But then the superiors that he had trusted had given him orders that were not as clear. Before he knew it, he wasn't just killing terrorists and combatants. There had been many innocent people as well. And he liked it.

Like the mother and baby in the minivan...September 9th?

But now, he was revolted at the thought of so callously enjoying work that made civilians tools. He had told David that he had hoped remembering would free him and lift a burden he carried around. He thought he would be rid of an albatross. Instead, he remembered all of his sins and nothing personal about his life. It was as if he had accessed only half of his personal memories — the bad half.

Were there any good memories? he wondered.

Burns suddenly felt betrayed by his superiors.

This won't stand. No. I don't have to be that guy any more! I can chose my own road, he thought to himself.

It was the best thought he had remembered in a long time.

As perplexed as Burns was by these emotions, he followed his target carefully while he kept enough people between them so he could remain invisible.

By the time the attendant got to the second block, he was on the phone again. This time, however, he was more relaxed and slowed his pace. Then he stopped, looked around, and entered a liquor store. Burns suddenly turned to look inside a store window as if he was just another person shopping for something. He waited patiently outside. Burns wondered if he had been identified by the attendant and if his target was now slipping out the back to get the drop on him. But then the attendant was out of the liquor store with what appeared to be wine. Another thirty feet, and he was picking up flowers.

Are you really heading to a date? he wondered.

Burns watched with envious eyes. He did have memories of being with women, but he had no memories of actually getting "the stuff," such as wine, roses, and chocolates, that went with it. Burns found himself distracted as he wondered why he felt as if there had been women in his life before, sex but no closeness.

Wait a minute. Foley was a friend. Maybe there are some good memories in there, he reminded himself.

Caulfield's therapy was working as evidenced by him experiencing feelings of sadness, regret, as well as jealousy and anger for this man he followed, while pain for himself. As Burns focused back on his task of following the target, a thought about why he felt the way he did materialized.

With the exception of Foley, the other women were not relationships, Burns sadly thought.

This was a distinction Burns could make now but hadn't in the past. In the past, all people were objects rather than something meaningful. What "meaningful" meant was still unclear to him. It didn't take a brain surgeon to know that Burns's thinking was profoundly changing.

Was it the brain injury or treatment? Burns thought to himself as he carefully maintained distance from his target.

"The doctor knows what he's doing," Burns said quietly.

The attendant was back on the phone, juggling his

packages and searching for his keys as he entered the city's enclosed garage.

Who the hell are you talking to? You look pretty relaxed and distracted.

Shifting back into hunter mode, Burns knew that his next steps at remaining invisible were going to be tricky because the camouflage of people would soon be gone. He could only hope the target's attention would remain distracted. Burns made a subtle but important change in how he viewed this man. The attendant was not the same guy who had walked him to his appointment about an hour ago, nor was he the same person who had asked Burns about his day and said, "Why do you talk to a shrink anyway?" He no longer saw the attendant as a person – he now saw him as a target, a key piece, an access point to get more critical information.

"'Objectification of people' – an active process critical to training for high-risk field training, hostage negotiation, and spec-ops," he recited in his head from his first field training a million years ago.

The target was fully engrossed in his conversation and did not notice Burns above him in the stairwell. There were other people, but they were thinning out as Burns followed his target to the lowest floor. Burns closed the gap between them as the target finished his conversation and opened the back door of his car to deposit his cache.

"So are you looking for me?" Burns asked matter-of-factually.

The target suddenly turned, startled, already going for something in his pocket. For Burns, everything sped up and then slowed down. As the target pulled out a collapsible baton and raised it to strike Burns on the head, Burns charged forward, slamming and pinning the target to the door, well within the baton's effective strike zone. Burns hit him so hard that the baton fell out of his hand. As the target attempted to get his arms around his neck, Burns's left arm braced the target's neck while his right was already planted on the opposing jaw. As suddenly as the struggle started, it stopped. Burns applied a few pounds of pressure and forced the target's neck and jaw in two different directions. The target's neck snapped, and his body went limp. Burns watched the body slump.

Burns was distracted for a moment.

Damn it! I didn't want that to happen. It happened so fast.

He had not wanted to kill the man; he had wanted information. His reaction was so fast, purposeful, and deadly it caught him off guard. With full knowledge now that he had killed before, this bothered him — not that he had lost access to intelligence but that he could kill so quickly and effectively without thinking.

What the hell is wrong with you? You've done this before. Why are you feeling...bad?

His scalp started to itch. The itching seemed to be focused near his scars on his head, his hands, and parts of his arms. The itching was not unbearable, just noticeable.

Burns returned and looked around for any witnesses. There were none. He popped the trunk and picked up the body and then put it in the trunk. Next to the body were two suitcases, a briefcase, and a duffel bag. Burns looked in the duffel bag first — dark clothes, gloves, and an array of military fatigues and service coveralls. A smaller bag was in the larger bag, and it was filled with cash, credit cards, and genuine American passports that were blank, not to mention a few Canadian, British, and Irish passports as well. These American passports and all the others would be priceless in the right hands on the black market. As far as this stash of cash, there had to be at least thirty thousand dollars.

All standard-issue, deep-cover field agent. Maybe supervisory level with all this stuff. What else do you have?

One of the suitcases was not a suitcase at all but an oversized, impact-resistant case with a cache of guns and knives. On the first of two levels, there were two large double-action automatic handguns. The smaller semiautomatic would be easier to conceal than the larger semiautomatic handgun. There was also one high-caliber revolver. All the guns were well maintained and meticulously placed in their padded holders. On the second level, there was ammunition for all the guns and two sets of knives. One set was spring-loaded from the side, and the other set was filled with throwing blades. The spring-assisted knives were heavy-duty, and each had points on the handles to shatter windows and cut seat belts. All the weapons were military-grade, but there were no serial numbers or identifying manufacture information on either guns or knives.

Definitely a supervisor. I bet you have a stash of money for bribes and payoffs in one of these other bags. And all this hardware and material is for a team...I wonder where your friends are, he wondered as he collected both data and his thoughts for next steps. Burns took one of the spring-assisted knives and some ammunition for the revolver he now held. Though he initially chose the easier-to-conceal, semiautomatic weapon, he changed his mind for the more reliable revolver with

two burn phones were probably all right because they w
the target's own disposable phones. The other data includ᠎
files — files on him, the nurse named Samantha Littleton,
and Dr. David Caulfield. He knew the nurse's picture, but
there was very little data in the file. They were remarkably
basic — date of graduation, transcripts, postgraduate assignments, last employment. Data prior to college and nursing
school enrollment appeared to be juvenile detention records
from age eleven years old to age seventeen, and then nothing. Some questions about possible illegal activities, possibly
prostitution, but nothing substantial. After her first year at
the veterans' hospital, her electronic trail vanished.

Wow, you really are a mystery. This is all they got on you? How did you manage to remain off-grid? He couldn't get any- thing else? I'm impressed.

David's file, on the other hand, was thick and filled with a great deal of data, including a picture of him holding a woman's hand near a residential backdrop. He was smiling and relaxed. It must have been recent, and at his home.

"This is your wife," Burns said out loud.

David had spoken of his wife very rarely, but it was clear he cared for her.

Suddenly, Burns noticed a smell, a plastic but metallic smell, one that was different from the soldering smell he had experienced in David's office about sixty minutes ago. Burns looked in the backseat and saw a blanket. Reaching back, he lifted the blanket carefully and discovered a narrow piece of wood laid out neatly beside the wine and flowers. There were three spaces hollowed out with two of the spaces holding two blocks of explosives. Narrowing his eyes he did an

the most stopping power. He eventually decided to take the smaller semiautomatic weapon as well for backup.

The other suitcase had what appeared to be the target's clothes — business casual and silk shirts and shorts for somewhere tropical. Another section of this suitcase held about fifty thousand dollars neatly arranged in a concealed compartment.

Ah yes. There you are? Daniels has a whole team on me. The must be covering Dr. Caulfield too. I'm guessing the nurse is al ready hiding. She seems like she's smart enough to stay low, h hoped.

All the cash was in small denominations of tens a twenty-dollar bills. Burns closed this up, took the briefc out of the trunk, and placed it on the ground as he searcl the body for the wallet, keys, cell, and any other thing of telligence value. Once found, he shut the car trunk with body and luggage and sat behind the wheel of the dece man's car. He then opened the briefcase on the passe seat of the car. The first piece of critical data was a that granted the deceased target access to a military ai Burns had seen this before — *no customs, no checkpoint no security. This guy was going with everything in his tru questions asked.*

There were two cell phones — burn phones tha undoubtedly purchased to be disposable later. There v a small laptop. Burns decided he would open the lapt when he could look at its contents and then leave it assuming that the moment it was turned on, it would s its GPS location. He made a similar decision regar target's personal phone; he would wait to open that l immediate risk assessment: *military explosives, not the commercial mining brand; wired for detonation with triggers right next to them... and one set is missing.*

Burns carefully got out of the car to make sure it was not the car that was rigged to explode. Convinced it was not his new ride, Burns began to wonder where the charge was.

It only took a second for it to become clear — *David Caulfield.*

Is it in his car or office or home?

Feeling a sense of immediacy, guilt and anxiety – mostly feelings he was not familiar with - Burns felt compelled to get back to David's car and make sure it was not rigged to blow up. Why he wanted to ensure David's safety and even warn the nurse he saw months ago was beyond him.

Burns started the car and drove quickly. He gave the garage attendant the ticket and money and drove back toward David's building. A few blocks later he saw David driving out of the parking lot while the two federal officers carried crates of papers and files. Burns drove by them and followed David at a distance so as not to be discovered. As he drove, he pulled a map out of the glove compartment to get an idea of where he was. The map was already marked with a highlighted route he was now following. It was still even clearer. The route led to David's home.

Damn it! They obviously know where he lives, but why wait to blow his car up? Why not just do it when he got in? Either remotely or trigger?

Because the size of the explosive was too small for an entire house, Burns reasoned it would be perfect for a car. The explosive was either in David's car or somewhere similar.

Then the hard part came: he needed to follow, watch, and wait while he used his former therapist as bait to lure out the other target. He was troubled by this thought and approach.

Damn it. What's wrong with you? Using a person for bait is not supposed to bother me. Collateral damage should be expected.

But for some reason, the fact that it was the doctor who had helped him bothered him greatly. Intellectually, this was something he knew he should not and had not cared about before. His ability to kill an adversary did not bother him; somehow, he was making a distinction between who could be killed and who should not be killed.

"Innocent people aren't objects but have more meaning," Burns said aloud so as to believe what he heard.

Maybe that's the distinction: my target was far from innocent, laced with malice. Caulfield, that nurse, Samantha — they're something more meaningful. They didn't have to do anything for me. They're something positive, Burns reasoned.

Burns drove in silence.

Chapter 7

"Nemo repente fuit turpissimus"
"No one ever became thoroughly bad in one step,"

– Juvenal

Four Years, Six Months Earlier

THE INTRUDER STOOD QUIETLY in the large laundry room in the basement of the hotel. Walking in tight circles, she found it helped her focus when she was on the phone. It was hard for her to really focus as the conversation was more social than she was used to.

"No, French food would be great. I've never had it before," she said as she tried her best to soften her voice and sound interested.

Now I have to look up what French food is. Is it that snails crap? I hate having to do this! Why do people do this shit, she thought as she listened carefully so as to not miss anything, any hint of how she was supposed act on her first date.

Who are you kidding. He's your only real date as an adult. Paid sex doesn't count. It's easier than this!

Suddenly, the thought of "paid sex" reminded her of what she had been looking for before she was interrupted.

"OK...I'll see you in a half hour at the rally point. I'm almost done here," she said as she leaned into a large laundry bin to make sure the maid was dead. After confirming there was no pulse, the intruder looked closely at the woman's misshapen arm and saw that the bruise she inflicted to make her talk was impressive.

No wonder she cried. That had to hurt.

"Ah yes. I like wine. Whatever you think will go with dinner," she said. Moving the phone to her chin and shoulder, she now was able to use both her hands in extracting the maid's keys, badge, and walkie-talkie. While still listening, she returned to searching the woman's cleaning cart for the list of rooms that were not cleaned, in the hope that it might give a hint of who might have remained in their room all day. Coming across the maid's small purse, she found a picture of the woman's family. Comparing the picture to the many others she had collected over the years, she found herself wondering exactly where in her collection she would put it.

Hmm. Nice picture. These kids are way cuter than many of the other ones, she thought as she carefully put the picture in her own pocket and finished emptying the purse of money. Finding the clipboard at last, she found there were ten rooms that were not cleaned. But there was only one at the far end of the building on the third floor that looked promising.

Well, well, well. I bet that if I was a whore making money, that's where I'd be.

Catching the last part of the conversation, she finally smiled as she responded to a confirmation.

"Sounds great. I just can't wait," she said with as much enthusiasm as she could muster. Closing out her line, she took in a deep breath as she tossed the clipboard on top of the maid's body and covered her with dirty sheets and towels.

Adjusting her newly acquired walkie-talkie to her belt, she walked quietly up the stairs to the third floor. Feeling for her collapsible baton, she found herself not looking forward to her date.

"I hate people. So why the hell am I going on a date with this guy? Because George dared you too, you moron," she said to herself as she crested the third floor landing.

You'd think I was twelve doing that. I don't even like men. I don't even really care for women either, she thought.

Stopping to look for the numbers, she found herself trying to figure out who she liked after twenty-eight years of living.

Goldfish! They're my favorite people. Better than people. Feed them, clean out their bowl once in awhile, and when you're done, drop them in the toilet. People are much worse to get rid of. Especially the women, crying and screaming, she thought as she stood in front of her target's door. Feeling for her baton again and planting her rear foot securely to the floor to kick the door open once it was unlatched, she knocked on the door as if she were security. As she heard steps closing in, she adjusted her newly acquired her security clothes.

"I wonder if she's as pretty as her pictures," she said quietly as she waited for her target, a nurse who worked as a hooker, came to answer the door.

Samantha appreciated the warning, and liked David

very much. He had never judged her, and he had gone out of his way to help her go to school and get another profession without asking for anything in return. This assistance without anything in return both confused her and drew her to him. He was a man but did not want her for sex. She knew David was straight, and she liked his wife. Samantha was genuinely appreciative to know David. He had opened her eyes to other possibilities.

Maybe if there were more people like him than the assholes I know, she would often think to herself.

Unfortunately, he helped her access feelings of remorse, regret, and sadness. While he and other therapists had given her tools to cope with them, her past abilities to compartmentalize her feelings and not feel as much pain had been greatly reduced. She did find that she loved her foster sister, Becky, more though. She even cared about her foster brother, Tony, as well, when she hadn't before. These feelings were good, but they were few. Even going back to prostitution was difficult when it used to be simply a way of making money.

Jesus, David. You almost killed my livelihood with all the self-analysis and getting in touch with feelings and shit. I wish I could have continued, though.

Nursing was a good choice for a legitimate profession, but it just didn't pay well. Still, Samantha was holding onto to her new dream: make a hundred thousand dollars in three years and buy a house in Willamette Valley in Oregon.

Low cost of living. Nice people. Beautiful gardens. I love gardens. I wonder if I'd be good at gardening. What does it really take? Maybe Becky will move out. Maybe even Tony if he's not being an asshole. This gig will get me there. After that, hello ER work.

She would miss David. He gave her hope and now a dream.

Samantha started to focus on her next steps. Instead of canceling her date set two hours from now, she decided she would get dressed, pack, check out of the hotel, and call him from the road after she was gone.

She was deep in thought when a knock at the door drew her attention.

A woman's voice on the other side announced, "Hotel security."

This is odd, Samantha thought.

Samantha threw on a robe that actually covered rather than "enhanced" her body. As she looked out the peephole and saw a woman who looked annoyed, she found herself feeling naked.

Well maybe that's because you almost are.

The woman was dressed in a security blue blazer, white blouse, and gray pants, and she was holding a radio. Samantha's head started whirring. Her last date had left thirty minutes ago.

He was loud, but not that much to draw that much attention. Samantha opened the door just enough to show the security officer that she was "just getting out of the shower, and could she give her a few minutes?" and suddenly, the door burst opened, knocking her onto the floor. As she landed on her side she looked up to see the intruder close and lock the door behind her.

"What the hell!" she yelled out as she tried to crawl away.

The intruder was a woman, but she was well-built — broad shoulders, thick torso, very dark hair, and dark features. Just as suddenly, Samantha caught the flash of something metal expanding from the intruder's right hand. As she continued to crawl away, she felt a sharp whack hit her buttocks, and cried out. Still in shock from the pain, she felt the woman pulling her hair upward to get her on her feet, forcing her body to follow. Then the woman yanked her hair in the opposite direction which sent Samantha hurtling to the bed.

Jesus Christ! Who the hell is this?! What's going on?

Face down on the bed, totally out of breath, she was violently turned over.

Looking into the intruder's eyes, she saw still darkness.

Oh shit! This is bad.

She had seen that look before, but never seen in a woman before, only men. The intruder straddled her, one hand holding Samantha's arms above her head.

"So, Danielle, or whoever you are," the intruder started, "I need some answers."

The intruder's violent behaviors did not match her low, almost calm tone as she spoke. Samantha found it very disconcerting, to say the least.

"What do you want? The money? It's in the drawer," Samantha offered.

"I will take that later. But before all that, do you know where Alex Burns is?"

Wow. This is real fucking bad.

David's warning had been right, and her gut feelings about getting out of town had been on target. It was much worse than she had thought. This intruder was not law enforcement; her methods were professional and to the point. There would be no paperwork, no hotel "incident report" done as one might have to do in any law enforcement agency when violence occurred.

"Are you kidding? I stopped that job two months ago. It didn't pay well," Samantha stated.

"I can see how you could make a lot of money in your present line of work." The intruder's free hand was now moving up Samantha's side of the opened robe and was now lightly touching her breast. Samantha immediately noticed that the intruder touched her more like the way a person would touch a foreign object rather than something pleasurable. Samantha was struggling to figure her assailant out.

What's going on…lightly touching me but she's so cold…

With no emotion on the intruder's face, Samantha tried to look back into her eyes to see if she was registering any kind of pleasure from touching her. The intruder's eyes snapped onto her eyes, and without any emotion, the intruder slapped her across the face.

Shit! That was a big mistake, she thought.

"Are you sure you don't know where he is?" the intruder continued in her monotone, professional voice with no emotion or any indication of what she was thinking.

Maybe looking in her eyes was challenging her dominance, she thought.

Samantha's stomach began to turn as the sting of the slap was fresh on her face. She knew this churning in her stomach was more about hatred and emotions brewing than fear and trepidation. She had had these feelings before.

Don't hit me, she thought as those feelings seemed to rise rapidly.

Samantha could see that some smell drew the intruder's attention. The intruder took a moment to sniff the air. Samantha had a sudden vision of a wolf when she saw this reaction. Still, the juxtaposition of her assailant's violent behavior seconds before and her asking questions in a nonchalant fashion was startling. Without looking into her eyes, the intruder asked her next question very calmly.

"Are you wearing perfume? What is it? It's familiar," the intruder discussed out loud.

"Your mother's?"

The question came out faster than she wanted. She wanted to strike back somehow. Samantha watched the intruder's lifeless eyes lock onto hers again, and then she slapped her while she held both of Samantha's hands above her head with just one hand.

Samantha was recovering from her last slap when she thought she heard the intruder say "You don't know her" without much emotion as she went back to touching her body.

Samantha's feelings of anger and hatred were escalating.

I hate being hit, she thought.

Trying to push the feelings down so she could try to think of a way out of her situation, she could feel old hatred, fear and fury build up, steadily, without hindrance. She really didn't want to do what she had done in the past. This was all so reminiscent of past violence.

No, no, no. I don't want to do that again...don't make me do that again. I'll kill you. I will have to kill you if you do that again, she kept thinking as her stomach started to more violently churn.

Throughout the encounter, Samantha had not really struggled at all in an attempt to convey that she was weak and helpless. It was easy to for her to pull from memories every survival technique she needed to get out of this mess. Past boyfriends and angry customers had provided her with years of experience. She had done this before as a way luring the person into letting go and giving her an opportunity to escape. But this was strange; this was out of character for a woman to be so determinedly violent without provocation.

What is your problem? What he hell is your story?

Regardless, this luring and feigning weakness would have to work if she was going to live. She now conjured up memories so that she would be able to sob. It was easy to come up with such a memory with her family history — the pain, distrust, violations, and betrayals. The tears started to flow as she began to whimper.

Samantha saw that the intruder stopped touching her and was looking at her face. Not her eyes, Samantha noticed.

She must be looking at the tears, she thought. The intruder wiped the tears off of Samantha's cheeks as one would brush salt off of a table. Calmly, her assailant spoke more to herself than to Samantha. It became clear to her that this woman saw her as an object. Maybe a lower form of life. *Maybe a pet.*

Samantha was sickened, feeling anger burn in her stomach.

I'm not an animal! You'd better not fucking hit me again! I'm not going to be a victim, Samantha's thoughts seethed.

"I have to go soon. I have a date later on. If you tell me where Burns is, I might let you go. It could be fun…"

The intruder's monotone, emotionless voice trailed off as she went back to exploring her body as if she were a specimen.

Samantha knew there would be no "fun" with this one. If the intruder had been a man, she might have had a chance. She could have offered sex or given him money. Then through the fake tears and whimpering, Samantha figured it out. Even as her fury built up to a crescendo, she remembered David's course in psychopathology.

She sounds like a guy. She dominates and seizes power like a guy. She's a sadist! An antisocial disordered sadist. She doesn't want sex – she wants me to beg. Screw you! You think I'm going to beg if you plan on killing me anyway? Fuck you!

David had always said, "It's a good thing the prevalence for this type of person is very small."

Finding her anger brimming, filled with past pain, her enmity was solidified, and her resolve focused as the same cold feeling she had experienced a few times before returned with vengeance. Her heart was racing. Her instincts were sharpening to strike back.

I'm not an animal! Thoughts repeated, thoughts screamed.

I'm not going to be a victim again, they continued as she felt her muscles begin to coil so she could strike back. Samantha went to a dark place. She had killed before. She didn't like to. She feared going to hell.

But this asshole is not going to fuck with me!

The intruder was busy feeling Samantha's body with her left hand while continuing to hold both of her hands above her head. But the intruder's grip was not as solid as she explored Samantha's body. Samantha continued to whimper. With the robe fully opened, the intruder had full view of her "working" outfit, which was thigh-high lace stockings, high heels, and a thong.

As the intruder's right hand started to move up inside her thigh, Samantha clamped her thighs shut, trapping the intruder's hand briefly. A low, guttural scream emerged from Samantha's throat, catching both her and the intruder off-guard. Both were enough to distract the woman as Samantha brought the weight of her two hands down on the intruder's face and knocked her off balance for just a moment. This allowed Samantha to swing her head and bend over the side of the bed. She knew she was not going to get off the bed with this move. She just wanted to get to the crevice between the bed and the wall above the floor.

That was the plan. As she hurled herself to the side of the bed, Samantha looked for her salvation, a fixed, double-edged knife that was nestled between the bed frame and mattress.

Hanging upside down, looking frantically, she finally saw it.

Thank God!

Clasping it for her life, she waited for the next move: she was planning on the intruder yanking her back on the bed. That would give the appearance that she was still a victim and weak, and it would also give her the much-needed momentum to use the knife. Samantha felt hatred and anger toward this woman for making her kill again.

I hate this! I hated them too. The ones before. You don't know me. You have no idea how much I hate you!

"Bitch!" she heard the intruder yell.

Finally, an emotion! Fuck you! Samantha's mind yelled.

Just as predicted, she felt a punch in her side that made her cry out for real. It was hard for Samantha to will her body to go limp. But she had to if she was going to kill the intruder. Samantha's mindset was honed on one thing – *kill her*. She didn't want her to survive the assault. That time had passed. She wanted to kill the woman. Samantha went limp as the intruder violently yanked her back into the middle of the bed. As she was pulled back, her left arm went wide, and it was caught firmly by the intruder. Samantha's right arm, however, vectored faster than the woman had expected and slid into the side of the intruder's head. She felt very little resistance on the knife's hilt as it entered the intruder's throat. Samantha knew she had hit an artery because she felt warm liquid shoot up onto her arm.

Jesus! she thought as she fought to get away from the blood shooting out. Samantha didn't need her nursing degree to know this wound would be fatal. The intruder knew it too.

"What!" the intruder groaned, both her hands clutching at her throat. Looking at her, Samantha could see that the intruder didn't know whether to pull the knife out or leave it in. As a result of the strike, the intruder lost her balance of dominance over her and fell on the bed. Samantha scrambled away from her and watched as the woman began to fade, waiting for the death gurgle to come. Trying not to look she watched as the intruder's hands clutched her throat to slow the bleeding down. Her breathing became erratic, and she gulped for air. In just a few short moments, the intruder's movements and her gasps for air slowed down. Eventually, there was a sound of a gurgle and a release of all muscles. Finally, the intruder's body was completely motionless and silent with her eyes open, looking at Samantha who was squatting against the wall. From a curled-up position hugging the wall farthest away from the bed, Samantha looked briefly into her eyes. The intruder's eyes seemed as lifeless as they were before.

Is she really dead? she wondered.

At first, Samantha covered her ears and watched passively. Samantha became consciously aware of her own voice muttering repeatedly to herself - "Why did you make me do it?"

The anger and hatred that had built up to Samantha stabbing the intruder were passing just as life ebbed from the woman. She felt empty but fearful.

"Why did you make me kill you?" she asked blankly, feeling suddenly devoid of emotion herself. When Samantha was sure that the intruder had passed away, she removed her hands and looked blankly beyond the dead body.

David had explained to her once that when people had been victims of violence and survived, they were more likely to experience the prior events of violence again. Samantha had asked him what helped to cure this, and he had said, "Treatment, love, work, more treatment, and time ... preferably in that order."

Samantha had seen death a number of times. This was the third time she had killed someone. Her step-cousin — or rather the cousin of her foster care sister — had been the first when she had been fourteen. The other had been a "date." The man had been in his car and had forced her to do fellatio while he had been holding a knife to her head. He was not aware that she had had her own blade much closer to an easier target. That one had been more difficult as the first two stabs did not kill him, so she had to stab him repeatedly in the back to make sure he was dead. She always vanished after. She would again. The visual images, smells, and sounds of the flashbacks began to recede when Samantha started to focus on her own breathing and stood up.

It's never easy to kill, she thought.

Still, this killing was different. The coldness and emptiness sat with her as she remained crouched against the wall.

Could I become as lifeless as her? she thought.

Samantha began to feel nauseous at the thought. Not only was she a woman, but this time, Samantha had killed with just one blow.

Maybe killing is getting easier for me, she reflected.

Another wave of nausea hit her.

Samantha had no idea how long she stayed there looking at her intruder, now victim, when she was startled by her phone ringing. She jumped, and her body exhaled when it rang. She did not answer. She eventually got up to see that the intruder was indeed dead and then took a shower. When she was in the shower, she threw up. Once out of the shower, she threw up again for what seemed to be ten minutes. The pain on her backside where she had been struck was pulsating now. Ibuprofen and ice was the answer, but she only had time for the ibuprofen. Her nursing degree allowed her to appreciate the impressive bruises forming.

"I am going to feel and look like shit in a couple of hours," she said quietly.

As calmly as she could, Samantha methodically pulled together all of her belongings, which didn't include much. A change into "civilian clothes," an inspection of the room to make sure nothing was left, and then she had to get her knife back. With a plastic shopping bag in hand, Samantha pulled the knife out of the dead body on the bed and wrapped it in another bag. The bed was now drenched with blood that pooled everywhere. Samantha then rifled through the body's pockets for any wallets, making sure not to touch the blood. There was a billfold of cash. Maybe about five hundred dollars in tens and twenties. In addition to this cash, there was an envelope containing about two thousand dollars. There was also a wallet with the identification of someone else — a chambermaid. There were also keys for the hotel, and obviously a picture of a family.

"Ah shit...she had a family. I didn't need to see this," she said as she gently put the picture back in the intruder's pocket.

Just great. You killed a mother...

Pushing her feelings aside and remembering that the intruder was going to kill her, she continued until she found a white substance in a plastic bag.

Still...what kind of mother has so much cash and drugs, and needs to kill people? For more money? It doesn't make sense, Samantha kept thinking.

While not a drug user, she knew what cocaine looked like.

Samantha made some more decisions. She would leave

the cocaine, wallet, billfold, and keys but take all the money. Samantha also took the dead woman's cell phone.

Samantha knew her prints were all over the room and that it would only be a couple of hours before this mess was discovered. With the air conditioner on high and the "Do Not Disturb" sign on the door, she hoped that might get the twelve hours she needed to disappear. Because she had no police record, there were no prints for law enforcement to compare her prints to.

The room and the rental car were under an alias, so Samantha's identity would be difficult to trace back to her. Fortunately, juvenile records were sealed and not available to law enforcement databases.

As she closed the door and started to walk toward the stairwell, her phone rang again. She answered in her working voice, "This is Danielle."

Even through everything she had gone through in the last hour, she was able to turn on her working voice and feign interest. It was her date, who explained he would be much later than he had anticipated. It was not difficult to fake disappointment but tell him that she would not be available until next week.

"Maybe next time," she said. She offered him a lower rate for the inconvenience. While disappointed, she could tell that the fee reduction seemed to make him happy.

Why can't I worry about getting a real date? Having a real life? Meeting a real person who cares about me? Why do I have to kill? Why do they make me do it? Samantha wondered.

She already knew the answer.

That's just not my life.

As the phone call ended, Samantha felt the warmth of the air on her skin. It was a beautiful night to vanish. Her plan was to call David once she was on an express bus out of the state.

Maybe tomorrow. He must be heading home now. But how can I call him if he dumps the phone like I told him to?

She was not about to call his office either. She did not care much for the office manager, Michele, and she was positive Michele did not like her.

If I leave a message, I'm sure Michele will misplace it. No, no...I'll simply leave and vanish, Samantha decided.

She knew she was going to miss him, almost as much as she missed Becky. It was at that point Samantha decided she would make an effort to see her foster-care sister. Samantha knew her sister genuinely loved her the same way Samantha had always hoped David cared for her.

Like family. A real family.

The bus depot smelled of old sweat and fast food. There was an express bus heading west. She decided to take that bus first and then head north. After a month, she would head back to the Northeast, where she had some connections and could easily reestablish another identity. Once she was well out of the area, she would dispose of the knife, clothes, and the intruder's cell after she deleted all its information. The SIM card would be thrown out in a different place.

Fortunately, the wait for the bus was only twenty minutes and the ride would run the whole night. With a croissant in her hand and a cup of black coffee in the other, a wave of fatigue came over her. It was about an hour into her bus ride when she felt as if she were falling asleep. There was still the pain from the assault and only now did she feel the bruising on her wrists. Her mind trailed off. Samantha felt sleep was coming on, but she didn't want to fall asleep until she was more than an two hours into her trip. She loved sleep, but letting go of control was difficult for her. Samantha was trying to push her violent attack out of her mind and think of her sister. In her sleepy state though, her memories of both merged as she recalled her cousin.

Becky was three years older than her, and she seemed to like Samantha from the moment she arrived. Samantha remembered feeling special for the first time in a long time. It seemed to her that Becky could read her like a book. She was only ten years old when she had arrived, and Becky seemed to be much older than thirteen. Even then, Samantha could easily see that Becky took care of her parents too. She was always cooking, cleaning, and repairing things while their parents worked any shifts they could at the plant. Samantha always wondered how Becky could take the time to help her with her school and why she actually liked to include her with her older friends. Getting out was rare though, because Becky had a lot to do, Samantha knew. She also knew that Becky made her brother, Tony, watch out for her too. But he didn't seem to like her very much. Samantha often wondered why.

Feeling her head drop as she dozed, she forced herself awake.

No! Not yet. Stay awake. I need to get far from here first.

She looked outside at the darkness and saw her reflection in the window. Her thoughts felt dark as she remembered Becky and Tony's cousin. He was four years older than her and had a knack for showing up when Samantha was alone, typically at night, usually around bedtime in the beginning. But then it seemed he would show up at any time no one was around. The sexual abuse started like it had at other homes: accidental touching or "playing doctor," or she would hear "Let me teach you about the facts of life."

Always some dumb excuse. All lies.

And then came the threats — if she told anyone, she would be hurt. After years of abuse, Samantha couldn't take it anymore. She loved Becky, but the fear was too much. Every time she told someone before at the other foster homes, they made her leave.

"I don't want to leave Becky. I like it here," she would say to herself so she could keep the secret longer. That's why she never told anyone. When she did tell Becky, her sister was pissed.

But not at me? It's not my fault, Samantha finally believed. After she told her, Samantha noticed that Becky watched out for her more. Tony got in a fight with the cousin shortly after Becky told him why she had wanted him to watch out for her. There were still times that Becky would get home late from band practice while her parents picked up overtime.

My foster parents never were around. Where were they? Why did they just dump everything on Becky and Tony? They were just kids too, Samantha often thought whenever she looked back objectively.

One of those times when she was home alone, the cousin showed up out of nowhere. As hard as Samantha tried, she couldn't recall how she'd found a knife and killed him. She remembered his warm breath, his hands all over her,

her pushing and running into the kitchen, and then she felt something warm on her hands and her feet slipping on the floor until she landed hard on her buttocks.

And those dead, lifeless eyes, Samantha remembered as if it had just happened yesterday.

It happened two hours ago. It just happened to someone else, she reminded herself.

Samantha could still see Becky just standing above her with a towel and her bathrobe. She heard bath water running as Becky walked her to the bathroom.

"Why did he make me do it?" Samantha could still hear herself saying those words all those years ago.

"You had to," Becky said in an even tone.

Her voice wasn't warm; nor was it judgmental. Almost flat. The warm red water seemed to ebb and flow while Samantha watched Becky going in and out of the bathroom to the kitchen. Samantha heard another voice outside of the bathroom. It was Tony's.

"Fuck, Becky! You really want to do this?" was all she could hear.

It seemed as if it took forever, but she finally was cleaned up. Samantha was frightened to go through the kitchen to get to her bedroom, but he was gone.

Everything looks normal. Like nothing happened, Samantha thought.

Samantha slept in Becky's bed for weeks after that. She hated sleep, but it felt all right with Becky beside her. After several days, Samantha felt bad for her sister. Becky seemed tired and drawn all the time. While she was home more and reading, she stopped going to after school practices, and cleaned the house, focusing on cleaning the floors and folding laundry very carefully. Samantha also noticed she was eating everything in the house.

She never did that before. I've never seen her eat like that before. Ever.

After weeks of watching her sister become quiet, withdrawn and getting bigger, it was easy to see that Becky was getting tired faster and would get winded quicker. This was strange since she had always been athletic and Samantha was just not use to it. As the weeks turned into months and Becky's condition got worse, Samantha offered to tell the truth and go to jail.

Becky looked at her as she closed her book, sat up on her bed, and held her firmly by the shoulders before she spoke. It was a calm but sad voice Samantha remembered. Not accusatory, but rather resigned.

"No, Pumpkin. You didn't do anything wrong. He did. When you screw with kids, you'd better be prepared to get fucked back."

That was the last time they ever spoke of it. For Samantha, compartmentalizing feelings and keeping secrets was as easy as breathing.

Not everyone is like me. It's nothing for me...but not everyone can do it. Cleaning up the murder of your cousin, even if he was a jerk, is still a big secret keep. It had to be hard for Becky to keep silent and pretend she knew nothing.

Samantha noticed she was now two hours into her trip. While it was an arbitrary time limit, she felt as if she could sleep now. It had taken years for her to like going to sleep. After years of vigilance and needing to be in control while maintaining her "self-protection mode," sleep was the only place she could actually rest. It was the only time she could be free from the world, a place where pain could not reach her.

Preparing for sleep and falling asleep were difficult for her, but once she let go, once she was asleep, it was the only time she felt safe. As she drifted into unconsciousness and lightness began to fill her head, she still held on to one cogent thought right before she surrendered to sleep.

I wonder how David is doing, she thought.

Chapter 8

"Omnia mors aequat"
"Death makes all things equal"

Present Day – May 2

ANDERSEN WAS LOOKING AT Coleridge and trying to determine if he believed what Coleridge was saying when he was startled by Jefferies knocking on the door and entering the interrogation room. Coleridge had just wrapped up his statements about how his former patient had left his office abruptly and he was now being interviewed by federal officers.

Andersen got up to meet Jefferies halfway, and much to his surprise, the man he knew as Coleridge did not move at all. Even with his back toward the door and not on his own turf, Coleridge seemed pretty comfortable in the hot seat.

Maybe this guy is used to dealing with these situations.

Jefferies was a young officer just back from four years of military service and just out of the police academy. He tended to be obnoxiously upbeat and positive most of the time. By the way he was looking at Coleridge, Jefferies seemed disturbed. Jefferies handed Andersen a note; it was more a fax than a brief note. Jefferies waited until Andersen gave him a nod indicating that he understood the meaning of the note. Jefferies turned and left as quietly as he had come in. His discomfort was obvious as Andersen read the paper yet again.

He stood reading the document for a full minute and then folded the paper. He took a moment, but then he walked to his chair and considered how he would share the information. Once at his chair, he remained standing with the folded piece of paper in his hand and slowly sat down.

Maybe I won't say anything. Maybe you can tell me. What the hell is going on here? What's your story?

Andersen felt tired. Rather than pulling his chair closer to the desk, he remained in his seat with his arms folded. Throughout this nearly silent interaction, Coleridge remained still, his hands on his lap and legs crossed. Andersen waited for a response. Coleridge eventually complied by saying something.

"Lieutenant? I noticed your breathing has changed rates and depth. It sounds slower and deeper as if you need more air. I can sense you are not sitting at the table and taking notes but rather sitting away from the desk. It sounded like you had a piece of paper, a note about something. You must have found something out that is very surprising or at least unexpected. And it concerns me," Coleridge speculated.

You think you know everything?

"Why do you think it's about you? Maybe it's about my brother who is in the hospital?" Andersen lied.

"No. The person who brought it to you said nothing,

and you remained silent. It was as if you didn't want me to hear. And since I am in the room, I assume it's about me," Coleridge concluded.

Much to Andersen's surprise, Coleridge added one more observation.

"The aftershave the young man wears clashes with his shaving cream."

If Coleridge could see Andersen, he would have been offended at how Andersen was staring at him and sizing him up.

"So how long have you been blind?" Andersen asked.

"Four years, six months," he answered blankly while turning his head slowly, and cocking his head as if to listen.

"That would be about right," Andersen calculated.

Coleridge returned his focus squarely on Andersen.

"So ... the note is about me in particular? I am guessing you just found out that I am dead. Not just the poet, but me, David Caulfield," he said with little to no affect that Andersen could discern.

Andersen read out loud the neatly written note that was about four years old: "Doctors David and Jennifer Caulfield were victims of a car explosion and fire. The consuming fire was possibly ignited by a fuel leak in their car touched off by some sort of spark. Due to the intensity of the flames, visual and dental identification was not possible. Identification of the bodies was made through surviving personal effects near the blaze: watch, bracelet, and general size and weight—"

Caulfield interrupted in a very low tone, "And a smart phone."

"What?" Andersen queried as he looked back at the note

that confirmed he was right. But it was one of the items Andersen had not mentioned.

Only someone there would have known that. Caulfield and Coleridge are the same person ... except Caulfield is supposed to be dead...just like the poet, Andersen concluded.

"My smart phone was in or near the wreckage as well. They were able to identify Jenny by some of her jewelry. My body, however, was never 'completely' identified," Caulfield summed up.

Again, Andersen saw no movement from his witness.

A body found in a trunk of an abandoned car the next street over was also mentioned in the report. While suspicious, the local police never got a chance to complete their own forensics and investigation as "national security types" took everything over and eventually whisked everything away while the local police were relegated to keeping people away from the crime scenes.

Andersen sat quietly, looking at his witness. He really didn't know what to make of him. Caulfield or Coleridge or whoever this guy was, he was in the center of this, and he knew a whole lot more than he was telling. Andersen began to rethink his entire approach to the interview. Because of Caulfield's visual disability, Andersen wondered if he was not pressing as hard as he should have been.

Is it because he's blind it's keeping me from thinking more critically? Maybe he's not a victim? Maybe he's responsible for the shooting directly?

Andersen had a myriad of thoughts running through his head when Caulfield's voice brought him back.

"Lieutenant?" Coleridge asked.

Andersen was surprised by the sudden change in demeanor and question. No longer still, Caulfield sat closer to the desk, leaning on his elbows, his hands folded below his chin. Not waiting for an answer, he asked, "Would it help our relationship if you started calling me David? You seem distressed."

Andersen thought that maybe the approach of having his witness "take care of him" might get him the answers he needed.

"Maybe," Andersen answered.

They were both quiet again for only a moment until Andersen started again.

"Okay…David," Andersen emphasized the name to confirm the new name to use when he was addressing Coleridge.

That's going to be tough.

"David? How did you survive and your wife get killed?" Andersen had to find this out. This might be the emotional leverage he needed to press David for more. It would also solve a cold case.

David sat back in his chair again. He was going back in time again. Andersen was getting used to this occurrence.

"Every time I think about that day, I re-experience my wife's death. Talk about trauma!" he said.

Andersen remained silent, but he moved his chair closer to the desk to take notes. He thought that might convey to David that he was poised to take notes, if David would just start talking.

With a deep sigh, David took in some air and then he started, "This could take a while."

"We got time," Andersen said as he looked at his watch.

Andersen watched as David sat up straight in his chair. It was obvious to Andersen that David was gathering himself for a very long and confusing story that would hopefully lead up to the events that had happened in that house.

"It wasn't just a bad day. It was a shitty day on a monumental level," David began.

Chapter 9

"Omnia mutantur, nihil interit"
"Everything changes, nothing perishes,"

– Ovid

Four Years, Six Months Earlier

"GOD! IT'S REALLY BEEN a shitty day," David said to himself as he finally got in his car to leave. He had spent relatively little quality time with the federal agents even though he took his time collecting Burns's patent notes, files, appointments, and even scribblings he did in his presence. Just as he was being advised by the federal officers that his cooperation was still required for the future and that "not leaving town" was more than a suggestion, David remembered it was the third Thursday of the month. Jenny was undoubtedly going to call him about getting home to meet Bob and Carol somewhere.

Great. Just great. Can anything else go wrong? Flat tire maybe?

David was still wondering what was going on. It didn't

take long for him to start thinking about his last client of the day though.

Alex remembered something important, David surmised.

David began to obsessively recall the last three sessions to see if there were any signs of suicidal or homicidal ideation or plans.

He's capable of hurting someone, David thought.

But he didn't identify any specific person, and he seemed more distressed by the memories and what they mean as a reflection of who he is, David continued to think.

"And where did the feds come from?" David asked out loud as he drove.

David knew that the world had lots of real problems and real conspiracies going on, but he always figured he was a "nobody" in the grander scheme of life. Still, after months of working with Burns, he liked him. Burns was smart and witty, and he seemed to have great insights. There were a lot of emotions though; they ran deep. David could see that Burns was also hungry to figure things out at any cost. David admired that about him.

With knowledge comes pain, David pondered.

He knew that was a phrase from somewhere but couldn't place it. David smiled suddenly.

If Alex was here, he would know, David guessed.

David hoped Burns was all right.

Maybe I'll call the hospital before we go out, David thought to himself.

The ride home for David was fortunately quiet. He found the radio too distracting when he was driving now, and while he was not a fan of his secluded home in the woods, he was looking forward to simply lying on the couch and watching a mindless show. Maybe a book. Seeing Bob and Carol was definitely not on his list for tonight; *drinking expensive beer and playing pool with Bob could work though.* Time ticked away as the miles passed, and David could not help but find his mind wandering to what Burns might be remembering and if Samantha was all right. These two people were "survivors," if there ever was such a word that described overcoming the impossible. David's mind wandered so much he nearly missed his exit. As he turned on to his street, he saw the familiar homes. His own home had lights on, and the lawn and bushes were finally cut. His house stood in stark contrast to the two houses on either side under construction and the house across the street that was fully gutted right down to its exterior walls.

As David closed the car door, he picked up the pace to get inside to his waiting wife. David was genuinely curious about what he would find on the other side of the door. Would he find a wife who would be happy to see him and want to go out, or would he find a wife who would be annoyed at him and dread going out at all? The front light turned on as he approached. Once the door opened, David deposited his jacket and briefcase in the foyer while he went through the mail. Jenny walked toward him with a smile. She was wearing a new dress with matching shoes and a clutch. She took his keys from his hand and kissed him briefly as she passed by him to get to the car. David noticed the dress had a matching wrap as well. The entire ensemble was very flattering, but it did indicate that they were going out to dinner at a nice place. But for David, it was a win. He was with his happy wife. David dropped the mail he had collected on the way in and followed Jenny to the car. She had the lead on him and went right to the front seat.

Still though, David's mood was suddenly darkening with worry. Going out didn't seem right. Not because he didn't want to. There just seemed to be a lot of loose ends. Loose ends bothered David.

I hate not knowing. What's going on? he kept thinking until he heard Jenny talking to him.

"I'll drive," Jenny said. And then she added, "You know how Carol gets when we are late, and you drive slowly."

Jenny was right. After a day like today, he really didn't want to drive and he really was not in a rush to see Bob and Carol.

"Oh, please," David responded.

David had his hand on the passenger door when he suddenly remembered he had Samantha's cell phone.

Damn it, David berated himself.

It's a good thing covert operations is not my full-time job!

As he took the cell phone out to make sure he still had it, he wondered where his own phone was. He felt his other pockets and produced his smart phone, but now it became clear he did not have his wallet. David was hit with several thoughts all within a second.

I still have a phone I shouldn't have, and if the agents find out about it, I could be screwed. Did I already commit a federal offense by omitting the truth about the phone and warning Samantha? Where is Burns, and is he okay? Why did he think Samantha was in danger? Is she?

David calmed his thoughts for just a moment and began

to realize that maybe going out to dinner was not a good idea with all these things happening.

More questions than answers, David thought.

David turned suddenly and started to walk back to the house.

"I left my wallet. See what happens when you rush me?" he said.

It was the only thing he could think to say so he could come up with an excuse to not go out.

The truth, maybe, he thought.

Jenny called back, "Don't worry. I've got money. Come on. I'll buy."

David thought how adorable that sounded. As he walked back to the car, there was a very loud crash that startled him, and did the same to Jenny because she was looking in the direction of the noise. It sounded like a ladder or something had fallen at the house across the street. David stopped in his tracks and peered at the house again and then turned back to Jenny, who was now looking at him.

He was in the middle of saying his last words to his wife when David's eyes were overwhelmed by a bright light.

David started to ask, "Was someone working at the Davidian's house—" when the light hit him.

Next, David felt as if he had been lifted off the ground as an enormous explosion filled David's ears. For a split second, he wondered if his house had exploded because the sound was so loud. He then felt his stomach turn inside out the way it did on roller coasters. He held onto this sensation until he felt the ground rush up to his back.

Jenny...are you...are you alright? I can't see you...I can't hear you, he thought as exhaustion seemed to grab him, and his breath was pulled out of him. The next sensation was warmth all over his face and chest. Then the world receded.

Not so bad, he thought. *Maybe I'm dying. Maybe I'm already dead.*

Fear suddenly grabbed him.

But where's Jenny? Is she okay? Why can't I move?

More thoughts flooded faster than he could process. There was stillness — no sound and no sight, just pressure on his back and a warm liquid on his face and chest … maybe arms and legs too.

Is this really it? Will I see Jenny again? Maybe she made it. Maybe the car protected her. I bet it did. She must have been protected…Maybe this is really it, he kept thinking.

In the silence, he felt still. A moment later, he had the sensation of someone going through his clothes and then the painful experience of being lifted and laying on his chest in a vertical fashion.

Shit … there's the pain. Am I sitting up? Is Jenny trying to help me? I think my eyes are open, he thought.

David felt drained and disoriented. He tried speaking, but his mouth felt so tired. He wanted to ask for Jenny and tried to move his hands to touch her. His limbs seemed lifeless but experienced his body being moved. The movements he experienced were not smooth and felt as if he was being carried until finally and thankfully was laid down on his back on a bed or something. He felt movement again, but it was more even and without pain. He had to be in an ambulance now.

Lying flat on his back, David still could not hear or see,

but he could feel burning pins in his body like sharp shards. Then the pain faded again until he thought he heard an echo of an explosion again. This one, though, was somehow muted. Suddenly, he felt really tired, and he felt like drifting off again. David made a conscious effort to make sure his eyes were open so he could at least see where he was. He could swear his eyes were wide open, but he saw nothing. In the darkness, he began to think of Jenny again until he felt so tired he couldn't stay awake anymore.

Maybe it was all bad dream. I know Jenny made it. I was outside of the car and had no protection. She was...safe...the car protected her...

Burns continued driving by David's street as David continued down his. Burns parked one street over. Needing to move fast though, he exited the car with his gun and knife but then turned back and took one of the high explosives and triggers. After he made sure the triggering mechanism was locked down to keep it from prematurely blowing up, Burns made his way through the sparse woods between neighborhoods and saw four houses. Three were in varying degrees of construction and renovations. One was clearly inhabited and well cared for as indicated by a cut lawn and trimmed bushes.

No guess work to figure out where he lives. And it's pretty isolated to boot. Perfect for a kill, Burns thought as he looked at the entire landscape for places he would hide if he were going to kill someone there.

Burns maneuvered to the closest house next to David's home while he kept his line of vision on anything near the house. It was easy for him to hear his therapist exit his own car, walk up the walkway to the front door, and go inside. The house light flickered on prior his entering the house. As Burns peered out to survey the area around David's house, he saw something familiar yet out of place - he saw the same car that had been parked in front of David's office, and it was now parked at the house across the street. Burns began to calculate how he would get across without being seen when the front door of the house opened again and a middle-aged woman came out to the car. She was well dressed for an evening out. David came out right behind her and locked the door.

Damn it! I need more time to find the guy...

Burns heard the woman say, "I'll drive. You know how Carol gets when we are late, and you drive slowly."

"Oh, please" was David's response.

Burns continued carefully scanning the area looking for owner of the parked car. Then there was a crash by the house where the target's car was parked. A ladder had fallen. Burns saw movement near the house and then turned back to see that David had stopped in his tracks. He was turning to say something to the woman in the car when it exploded. Burns found the explosion startling, even though he suspected something like this was going to happen. He instinctively shielded his eyes but turned back as quickly as possible to see what had happened and how big the crater was. The force of the explosion shattered the windows of the houses in the neighborhood. David was thrown back several feet in the opposite direction.

Burns resisted the impulse to immediately run out and assist David. Instead, he waited. Burns found waiting difficult. It was a new experience, too.

What is your problem? How many times have you done this – use someone as bait?

It was not long before a new target came out of the shadows of the house across the street and started to walk slowly and carefully toward David, who seemed to be moving slightly. His target, who had also seen movement from David, dropped into a crouched position as he moved. He drew a gun and aimed at the barely moving therapist.

Burns had to make a decision — either let the target kill David or kill the target first. In a past life, Burns would have simply let the target kill his objective. Then his target would relax and leave, Burns would simply follow. That should have been the logical plan. But Burns took another approach, one that was driven out of emotion rather than the cold, rational logic he was known for in critical missions.

Not today. It looks like you caught me on a bad day. Good for David but bad for you.

Completely out of character, Burns emerged from the edge of the neighboring house, aimed, and shot the target directly in the side of the shoulder before he could harm David any further. The target was thrown off balance. Two quick shots in the chest, and Burns was positive that his target was dead before he hit the ground.

Burns made his way around the fire and found David, who was either dead or unconscious. Glass and metal peppered his body and face. Though not lethal, the cuts on the face and near the eyes would have to be addressed quickly.

He checked David's vitals. They were there.

Wow. For a therapist you're pretty tough, he thought.

While the blood on his face and chest were worrisome,

Burns was really worried about a concussion and possibly eyesight as he did a rapid battlefield assessment of David's injuries.

"Hearing might be compromised and you will definitely have scars if we don't get you to a hospital...but your eyes... That could be a real problem," Burns said aloud.

Making a mental checklist of first-aid supplies he needed and where he might get it, he made the difficult decision not to bring David to the hospital.

Sorry David. If I bring you there, they will kill you.

Burns rifled through his pockets and produced a smart phone. Next, he took David's watch, bracelet, and pen and threw them near the burning car. Burns was about to throw the cell phone into the fire too, but then he decided to keep it. It was a burn phone. The smart phone was David's, so this phone was neither.

Maybe it's important, he reasoned.

Burns slid it into his pocket. He then turned and walked to the man he had shot. He picked him up, carried him, got as close as he could to the flames, and threw the body into the burning car. Burns gently tossed the explosive he had taken with him where he had thrown the body as well. Then he picked David up and carried him to the target's car across the street. Burns was aware that David's injuries were pressing on his shoulders and back and that they had to hurt David.

Trust me. I know what it's like to feel burns but there's no choice.

Finally, Burns made it to the car and was able to lay David on the backseat.

That has to be more comfortable for you, Burns speculated.

Burns entered the car as well, took the wheel, and started to drive out. Once at the top of the street, he triggered the explosive and created a second explosion. This would make the recovery of any bodies next to impossible. Hopefully, they would think the Caulfields were dead, and whoever had sent the hit men may think their guy had gone underground after he had completed the job.

Burns pulled up to his first car, which he had parked on the adjacent street. He got out and casually pulled the duffel bag luggage from the first car's trunk and placed it in his new ride. Burns carefully pulled away as all the witness's eyes were on the burning inferno the next street over.

For Burns, this whole affair seemed normal with one exception: he had no memory of ever saving someone's life. This was new. It was not a bad feeling but a new one. Then he thought about it some more. While Burns acknowledged that he was actually happy to have saved David's life, he also regretted not being able to save his wife. Burns was sure that this happy feeling and the regret associated with his actions were rare at best.

Samantha awoke with a sharp intake of air. It took her a moment to orient herself and remember that she was on a bus heading west. She also felt pain. Her wrists hurt but nothing like her backside. She had disposed of the knife, cell, and other materials at one of the bus stops. She had had an opportunity to get Wi-Fi and powered up her laptop. Her plan was to go to Madison, Wisconsin. She had been there before, and the people were nice, the cost of living low. Finding employment would be easier than on the East Coast. Becky might come and visit her there as well. It would be nice to see a friendly face. Samantha logged on to her "escort page" site to check the boards in the area when she noticed private messages for Amber. The first one was about a week old, and there were three recent ones in rapid successions. Samantha's heart dropped. Samantha and her sister, Becky, always had a communication system in place: if the phone numbers changed, Becky would contact Samantha's work e-mail, which was on this escort site. Because Amber was the middle name Becky had given her, it made sense for her to contact her this way. If Samantha had to get a hold of Becky, there were at least two e-mail addresses, a home phone, and a cell phone she could use to contact her. Since Becky was a paralegal and lived close to her parents just outside of Lansing, Ohio, finding Becky would be less of an issue.

Becky's first e-mail read, "Hi, kiddo! Everything is all right, but I need some advice about Tony. The number is the same so call. Luv, B!" The second e-mail came three days after and was dire: "Heading to Syracuse, New York, to that college you got in but never went to. Things are bad." The last one had been sent last night: "Will have to move soon. Call ASAP. Still in the town of the college."

What the hell? Something has to be bad for Becky to do this! What the hell did Tony do now?

Samantha got on her cell and called. The line picked up, but there was no answer. Barely able to breathe, Samantha asked, "Becky?"

A sharp intake of air, and then she heard her foster sister's voice.

"Thank God! Where have you been!?" Becky cried.

What!? What the hell? What's going on now?

"What's wrong?" Samantha pressed. The urgency in Becky's voice made Samantha forget about her immediate pain.

"Tony and his girlfriend are dead. His girlfriend was in some kind of trouble with the mob, and they're dead—" Becky started to cry more.

Tony's dead? What the...

"Becky ... I need you to tell me where you are and what happened." Samantha's ability to isolate her emotions during crisis was excellent in times of emergency. It helped when she worked the emergency room; it more than helped when prostituting. It did, however, keep her emotionally removed from people who loved her. That was a drawback she often thought about. Samantha frequently wondered what it might be like to feel real closeness. For now, Samantha pushed these thoughts out and focused on the crisis at hand.

What the hell are you thinking of this shit for? Becky needs you now!

Samantha noticed that Becky started to get control of her breath and started to pace it better. It took a minute and then she began very rapidly.

"I am at the hotel we stayed at when you looked at that college. I was going to leave today if I hadn't heard from you."

"What's going on?" Samantha said firmly.

"Tony had been seeing this girl for about eight months. She was pregnant with some other guy's kid she used to see. Tony was always a sucker for girls who needed help. They start dating. It starts getting serious. She has the kid, and he's really happy. One day, he brings the baby over for me to watch her right before I am on my way to work. I'm all pissed off, but I call in sick and watch the critter. I don't hear from him. I call his cell from my house, and after a while, I go to his apartment, where there are cops everywhere. I find out from the landlord's wife, who's crying and shit, that Tony and his girlfriend are dead and they are looking for the child. I head home, and there are some pretty mean-looking guys. Gotta be from the mob, so I go to a hotel and try to think. I turn on the news, and there's a picture of Tony and his girlfriend. The girlfriend's old boyfriend is part of the mob, and it's his kid. Now I got the mob and cops looking for me, so I panic and get out of Dodge. What am I going to do?" and then she started to sob.

Samantha's head was spinning. Here she was — she had just killed a woman, she had just left a former life that she liked, and now the only person in the world she knew who loved her was in trouble.

And just for shits and giggles, Becky's brother is dead, and there's a kid involved. Damn it! God must hate me!

"Are you there, Sam?" Becky pleaded.

"Becky, stay where you are. I'm in Springfield, and I have to change routes. If you need to go, go to the fancy hotel we stayed at in the city. There's a lot of people, and you should blend in. Okay?" Samantha asked.

Becky was still crying. She was older and had always had her head on straight. She stayed in school, and while she got into some trouble, she never got in too deep and stayed on the right side of the law. Becky loved Tony and he loved her. Samantha had to speculate and guess at Becky's pain. Losing him had to eat her up.

That's another advantage of not feeling, Samantha pondered.

Samantha liked Tony, and he would look out for her; however, it was more out of duty and love for his biological sister. He was a nice guy though and should not have been killed for doing the right thing.

"Becky! I need to know you heard me and know what to do," Samantha ordered.

That got her stepsister's attention.

She stopped crying for a moment and said, "Yes, I got the plan. When will you get here?"

"I'm on my way now. I'll let you know when I know. Love ya," Samantha added. Becky was the only person she had ever said that to and meant it.

"Love you. And be careful," and with that, Becky hung up.

"What the hell?!" Samantha said to herself quietly.

The events of nearly eight hours ago now seemed to dissolve in importance, and her world was now focused on getting to Becky.

By the time Samantha found the next place to change her ticket, time for departure, and destination arrival, she was able to call Becky back with an estimate. Samantha did change the plan. She had Becky meet her at a location in New York City — the Marriott Hotel in Central Square. This would shorten the time to get to her, and it would keep Becky on the move so that she would be distracted. Becky was a thinker, and waiting in a hotel any longer than she had to would drive her crazy. This plan solved two problems at once. Becky sounded stronger, and she was happy to be doing something proactive. Once she hung up, Samantha thought about what to do after she had arrived. While there were fewer people on this ride, the bus ride heading northeast was actually tenser than the one leaving Virginia. Once either she or Becky was about halfway to their destination, they would call the other — that was the plan at least. So when a call came on Samantha's cell only an hour into her trip, she was surprised and answered it quickly.

What!? Is something else wrong?

"Becky? Are you all right?" Samantha asked, closer to panic than she ever had been. There was silence for a moment. The line was open, but no one said anything. Was it a former customer looking for a good time?

"Hello?" Samantha pressed. She then looked at the number, and it looked familiar.

No way? Why would he call again? He was supposed to get rid of the phone? Something has to be really wrong for him to use it again.

"Dr. Caulfield?"

The silence broke on the other line.

"This is Alex Burns," the voice said calmly.

Samantha had never heard his voice as strong and as clear.

Burns? Alex Burns! The guy I helped out and now all this shit happened?

"Alex?" she asked. She was confused.

How did Burns get the phone I gave David? Something bad must have happened. Won't this shit ever end?

"Yes," Burns responded matter-of-factually.

That was when Samantha's confusion lifted and she felt the anger well up.

"'Yes?' Do you know I almost got killed because of you?"

There was no response at first, and then he asked, "Are you alone?"

Samantha looked over the chairs in front and behind her and lowered her voice.

"I am on a bus heading north, and I am lucky to be alive," Samantha added. She made sure not to mention where exactly she was headed. She really didn't trust anyone.

"I am sorry," she heard him say.

Samantha could hear something in his voice: *Softness? Sadness? Compassion? His apology did sound genuine.*

"Where are you headed?" Burns then asked.

"Why?" Samantha shot back. She believed it was Burns, but she was not about to risk her life or her sister's life on a gamble.

"Look. I have Dr. Caulfield with me, and he is hurt. And I need help."

My God, Samantha thought, *can this day get any worse?*

"Okay. Here's the deal. The past eight hours have been pretty eventful." And with that, Samantha summarized the events in Virginia with her assailant and murder victim, her trek heading out west, a wrinkle with her sister, and the time and general location of her final destination.

If Burns was either impressed or doubtful, he never conveyed it. He listened and then came up with a plan.

"All right. Once you get to your destination and secure the area, contact me, and I will be in the city as well. And we'll set up a meet."

"Is David with you?" Samantha asked.

"Yes. He is unconscious."

"What do you mean 'unconscious?'" Samantha asked.

It was Burns's turn to summarize the events of the past eight hours — his following the first target, following Caulfield to his home, the explosion, the killing of the second target, the money, guns, and clothes. He added that prior to leaving the state, he had located a somewhat secluded private medical practice, broke in and stole sedatives, antibiotics, and bandages. While he was able to clean and dress the wounds, it was clear that Caulfield might lose his eyesight. Getting emergency care from a hospital was not an option.

"So what the hell is going on?" Samantha finally asked. There was hesitation, slight but evident.

"We'll talk later. Suffice it to say that the people I used to work for are serious about cleaning up loose ends, and the sooner I get Dr. Caulfield to you, the better. I also think—"

"No. We're not going to 'talk later.' We're going to talk right now," Samantha interrupted in a hushed, angry tone.

"These 'people,' why do they want to kill me and David? Why did they kill David's wife?" she asked.

Samantha was cognizant of her surroundings. There weren't many passengers on the bus, and they seemed more interested in their reading, videos, and sleeping than paying her much attention.

"Because you know me. Because they might think I told you something."

Samantha could tell that Burns was used to short answers.

But my questions are pretty damn big, she thought.

Samantha knew a lot of con artists and scammers in her line of work.

Burns is a secret agent? A mysterious agency out for world domination? That shit doesn't happen to people like me or David, she thought.

Though she had to say that her attack, David's wife's death, and his injury wasn't just a coincidence.

Samantha decided to take another approach. She was going to assume his story might be true.

"Nothing personal, Burns, but couldn't they have just killed you a while ago?"

There was a short delay.

Trying to think up a good lie, she wondered.

"Okay. I know it's all hard to believe, but it's true. They could have killed me at any point, but they needed to know something first. They wanted to figure out something I did first. If I'm dead, they will never find out how I breached their relocation protocol and their cyber weak spot," Burns confessed.

"What?" Samantha was having a hard time understanding what Burns was saying. Her thoughts raced to Becky.

Samantha became aware of her fears building about her.

Is Becky in danger? Are these assholes after her too?

Samantha waited for Burns to start talking again.

This shit, it's too crazy to be bullshit, she concluded.

"You don't know these people. They just don't kill without reason. The guy who runs it, Eric Daniels, didn't create his empire by just killing people. He recruited talent. He used bright people to find holes in his systems, protocols, and defenses. He's known about me and has had everyone around me watched. He probably allowed me to be sent to David because he thought he could help me find memories. Memories Daniels wants. He had time to wait. I wasn't going anywhere. But if he just killed me, he would have to spend a lot of time and effort to find two key holes in his agency. Daniels didn't get to where he is by just killing everyone."

Samantha could actually hear stress in Burns's voice.

"How do you know so much about this guy?"

Who the hell are you, Burns?

Another short pause.

Samantha could tell Burns was not used to being asked so many questions, and he obviously didn't like giving up answers.

Too bad. Some people don't like being nearly killed, she thought sarcastically.

"I was on his senior team. I ran key operations. Operations so off-books that managers and directors had 'no need to know.' They knew what we wanted them to know. The true objectives, true mission goals, specific methods used, all were at my level and two tiers down. Daniels—"

Samantha noticed the pause.

Was he searching for a word to describe him? she wondered.

"Daniels was a colleague and my boss. We had a respect for each other. He may have wanted what's in my head before, but with me MIA, he's simply going to want to get rid of me and worry about the holes in his defenses later. That's why the sooner you get away from me, the safer you will be," Burns concluded.

"Oh? Like the nice woman who tried to kill me? You and I weren't exactly close, you know," Samantha said angrily. More silence.

"I need to put distance between you and me." Samantha's first image after Burns's response was Becky.

She's going to get killed because I tried to do you a favor? She doesn't deserve this, she fumed.

"Oh ... so we're slowing you down, Burns? Getting in the way of you crawling under a rock to hide while we fend for ourselves?" Samantha was gripping the phone with all her strength. The idea of her sister being in danger as a result of her was intolerable. With every word out of her mouth, she jabbed her finger in the air as if Burns's chest were right in front of her. Guilt was building up to an unfathomable point.

I already ruined David's life. I might as well have killed his wife myself. Fucking Sam! Why do you always screw up with people you care about? she thought to herself.

Then Samantha shook her head and tried to shake off her anger toward herself. She breathed in slowly as the phone line remained silent.

No. This isn't my fault. It's Burns's fault. He's just another guy who is screwing me over and making me feel bad. Maybe you can screw me over, but you're not going to do this to Becky. You're not going to do it to David, she thought.

Samantha felt a strange calm come over her. She pieced it all together. She never thought of herself as "book smart," and that was why she hadn't gone to Cornell University in Ithaca when she was accepted so many years ago. But she knew what Burns was doing. He was unloading his burden.

I'm not the asshole. He is.

"So that's the way it's going to be, huh? So you think you can just drop off the guy who helped you figure shit out and then pay him back with getting his wife killed, blinding him, and then taking away his whole life and just walk away?" She was pissed now. The other end of the line remained silent. Samantha felt her calm slipping rapidly away.

"Bad enough I have to kill someone, but I can deal with that. My sister is in trouble and needs me. Her brother is dead, and I'm the only one she has. I owe her everything, and now she might be killed because of you. Caulfield had a life and did me the favor of taking you on, even though he suspected it might be dangerous. Okay, how about you just fucking shoot the guy and end it quickly while he is still sedated. At least he won't live to see his whole life ruined!" Samantha regretted that part. It had come out faster than she had thought.

"I never asked you to help me," Burns said quietly.

I can't believe you said that, her thoughts yelled out. Her free hand balled up into a tight fist as she smacked it down on her thigh.

Damn it ... more bruises!

Samantha knew it was the anxiety about her sister, fear and guilt for David, and nearly being killed the night before that compelled her to say more things to Burns. If she had just thought about it, she probably would never have said what she said next. Her voice was low, harsh, and accusatory.

"Yeah, you're right. I just thought it was unfair that a guy with a head injury was being constantly sedated that I did everything to get him help. You know, Burns, I don't mind the fact that I fucked two administrators of the hospital to get you help. I do that for work. But if you think it's all right to ruin the guy's life who saved yours and then walk away, I hope they find you and kill you."

Samantha was quiet. She was shaking. She hated feeling

frightened. Her sister was in danger, an actual friend of hers was now completely cut off from his life, and the guy she had expressed compassion for was now walking away? If it wasn't for David and Tony, she could honestly say she hated all men more than ever.

Now Tony's gone, and David's life is over. All the other men suck.

Samantha was pulled back by Burns's voice.

"Call me when you secure the area." Then he hung up.

Samantha sat alone. She found herself absently folding the phone as she stared at the back of the empty seat in front of her. She couldn't believe how her entire life had changed from yesterday. How everyone she touched was now affected by her decision to help Burns. How different yesterday morning was.

"Yesterday was the easy day," Samantha muttered bitterly to herself.

Chapter 10

"A fronte praecipitium a tergo lupi"
"A precipice in front, wolves behind"

THE RIDE TO NEW York would be hours. It was hard for Burns to witness his former therapist deal with the loss of his wife; the disbelief, pain, shock and angst as David periodically talked to himself with choked voice. As time dragged on, David was now more clearheaded but only asked a few questions.

They were the questions Burns expected: "Was his wife really gone?" "Who would want him dead?" "Why her?" "Were his kids going to be safe?" His questions were all about others and their safety and not about his own. After hours of shaking his head in disbelief, gnashing his teeth and stifling tears, David fell silent for a long time.

Burns reviewed the conversation with his former nurse. She was right; killing David would have been the simplest and possibly more humane thing to do.

At another time, I might have done just that. Not out of compassion but out of convenience for me. Now the guy has to carry a lot of pain around...because of me. How the hell am I going to make this right?

frightened. Her sister was in danger, an actual friend of hers was now completely cut off from his life, and the guy she had expressed compassion for was now walking away? If it wasn't for David and Tony, she could honestly say she hated all men more than ever.

Now Tony's gone, and David's life is over. All the other men suck.

Samantha was pulled back by Burns's voice.

"Call me when you secure the area." Then he hung up.

Samantha sat alone. She found herself absently folding the phone as she stared at the back of the empty seat in front of her. She couldn't believe how her entire life had changed from yesterday. How everyone she touched was now affected by her decision to help Burns. How different yesterday morning was.

"Yesterday was the easy day," Samantha muttered bitterly to herself.

Chapter 10

"A fronte praecipitium a tergo lupi"
"A precipice in front, wolves behind"

THE RIDE TO NEW York would be hours. It was hard for Burns to witness his former therapist deal with the loss of his wife; the disbelief, pain, shock and angst as David periodically talked to himself with choked voice. As time dragged on, David was now more clearheaded but only asked a few questions.

They were the questions Burns expected: "Was his wife really gone?" "Who would want him dead?" "Why her?" "Were his kids going to be safe?" His questions were all about others and their safety and not about his own. After hours of shaking his head in disbelief, gnashing his teeth and stifling tears, David fell silent for a long time.

Burns reviewed the conversation with his former nurse. She was right; killing David would have been the simplest and possibly more humane thing to do.

At another time, I might have done just that. Not out of compassion but out of convenience for me. Now the guy has to carry a lot of pain around...because of me. How the hell am I going to make this right?

The problem was that Burns was having a real difficulty with just killing people who were innocent. That had never been a problem before, but now it was really affecting his judgment and behaviors. It was possibly endangering his new mission.

"Dr. Caulfield? This may be the absolute wrong time, but I have to know something," Burns started.

"Could you do me a favor and call me David? Somehow, I think we are now officially beyond the doctor-patient relationship," David responded in an empty, tired voice.

Burns was silent. He was trying to remember the last time, if ever, he had called someone by their first name. It seemed almost "too close" or "too personal."

David continued, "What's the question?"

Burns could tell that David needed questions more than anything right now. The thought of his wife being murdered right before his eyes had to be heart-wrenching. There it was again, Burns noticed. *Empathy? Sympathy?*

Burns pressed on with more questions. He could keep David distracted and get some answers about his new behaviors at the same time. More importantly, Burns needed to understand why he was having uncomfortable feelings.

"I am having a problem with killing people."

The statement, more than a question, sounded odd, flat, and almost anticlimactic, as if he was saying, "I am having a problem going to the bathroom."

After a long moment of silence, David cleared his throat and responded to the strange statement.

"By the nature of the question, I am assuming killing people is a matter of course. Kind of routine?" David asked.

Burns nodded and took a moment to organize his thoughts.

"I remember personally killing more than twenty-five targets. Maybe twenty-eight and even more collateral damage ... people around at the wrong place and wrong time."

Burns turned briefly to look at his former therapist. It was obvious that David was digesting the words as well as the volume of human destruction his former patient had committed. He did come back though pretty quickly.

"Okay. First thing I want is for you to promise me that you'll call me David. It is no longer about 'just getting to be friends' or to signify a change in our relationship, but I am hoping that if I have a more personal name like a first name, you might be less inclined to kill me."

Burns smiled.

Is David being funny or serious or both?

Burns caught his own smile, and was reminded that he had been smiling at small things — sights, jokes, conversations. This, too, was new.

Burns waited as he watched David trying to put technical terms into layman's language.

"Okay. While you may have killed and still can, there may have been a major change in your brain, which may have made it more difficult now. There was the initial trauma that may have damaged more of the limbic system, an ancient brain which feeds the capacity to feel anger, depression, anxiety, and the urge to kill. The limbic system is not all bad. It is all about survival; this is the 'old' part of the brain that is a holdover from evolution."

"So a head injury damaged my brain, and now I am feeling more stuff?" Burns asked.

Burns saw that David started to get up from lying down in the back seat but seemed to realize that it was a mistake and laid back down. Burns thought it was a good idea for David to continue lying down too.

David went on but it was not without pain.

"Not just that. Your limbic system may or may not have been damaged. I am guessing something happened to it. But I'm sure that the interventions you used to remember your past certainly affected another part of your brain. In particular, the treatment interventions focused on the higher order part of the brain called the frontal lobe. A very important part of the brain that makes us more social and civilized."

Burns remained silent. David picked up on the silence and did not wait for the obvious question.

"Exercise and physical activity probably unlocked the muscle memory, the kinesthetic memories. The meditation and learning new things probably significantly changed your frontal lobe, where logic, reason, empathy, sympathy, humor, love, and all the higher-order emotions and functions are. Your diet cleaned your body, and the medication that kept you sedated left your body entirely after twenty days."

It was finally becoming clear to Burns. His new awareness was a byproduct of David's treatment strategies.

Burns ventured, "So I developed the part of the brain that makes me—"

David finished, "—more pro-social. You now may be experiencing more positive emotions and thinking more positively, which in turn will affect your behavior. It's all connected. What have you been doing for meditation?"

Burns could see that David was now clinically curious

about his change which had been more profound than he had expected. As a result, it was easy for him to see the problem in his choice of meditation.

"Prayers," he responded quietly.

David repeated the word to make sure he had heard correctly. "Did you say 'prayers?'"

"Yes. The Lord's Prayer, Hail Mary in English and Latin," Burns confessed. Taking a moment longer, he gave more data.

"Eastern Orthodox prayers from the Orthodox liturgy in English and in Syrian and periodically Greek. I chose morning, midday, and evening prayers about thirty minutes each."

Looking back in the rear view mirror, Burns saw the academic curiosity that was evident all over David's face. He anticipated the next set of questions.

"How long have you been doing this?" he asked.

"Meditation was really the only thing I could start right off the bat, and the only things around at the hospital were various bibles." Burns stopped and then added in an almost guilty fashion, "So I started meditation sixteen weeks ago and started going to the chapel pretty regularly for an hour every other day."

David was quiet. Burns waited. He actually experienced anxiety as he waited for the next words out of David's mouth.

David managed to successfully sit up this time. Burns figured this had to be important for David to actually sit up so that he could give his diagnostic impression.

"Well, Alex, it looks like in addition to waking up your frontal lobe, you may have had a religious epiphany. Because of your choice of materials, which are prayers, it seems that the more sociopathic killer that made you an effective soldier has transformed you into a more moral, pro-social individual who values life. Sorry?"

Burns tried to determine if David's assessment was right, let alone a good or bad thing.

Is this bad news for all of us if my skills are hampered in this way? Maybe, he thought.

But Burns had to admit that he liked the fact that he was sitting and talking to someone about something important other than trying to complete a mission. That was another reason he enjoyed his sessions with David before all the shit went down.

This talking was different...nice, he had often thought.

It was new to him, having no prior memory of anything similar.

Burns reflected on how he watched his first target pick up flowers and wine for a date and his inability to remember doing anything similar in the past. Of past missions, he could recall a myriad of details.

Friends? Family? Loved ones...few to no memories. Maybe my mother for sure...

Burns felt embarrassed, another new feeling, when he stumbled over himself trying to describe his relationship with Eric I. Daniels to Samantha and he couldn't come up with any other descriptor than "colleague." Burns couldn't tell if he just couldn't remember anything personal about his life or if there was anything personal to remember at all.

Really? Colleague? Was that all I had? Was there anything more? Anybody? he thought.

Burns experienced an emotion that seemed familiar, but now he had the word to identify it: *sad. He felt sad.*

He shifted his focus to David and realized that he had more to say. With some noticeable pain, David continued.

"In addition, you voluntarily chose prayers of humility and forgiveness, which speaks to someone who may have been wanting to get out of the business of death and destruction but never had an opportunity until the crash and head injury. Kind of a 'Saul transforms to Paul' conversion. Are you still praying?"

Reluctantly, as if caught doing something illicit, Burns answered as quietly as he could as if not to draw any more attention to further shame him.

"Yes. Every day. Morning, midday, and night. When I try not to pray, I obsess about it until I do. After I pray, I can't tell if I feel better as a result of completing my routine or if I just love the words. I just can't tell anymore."

Burns realized that admitting that he did not have this answer had not been as painful as he had first thought it might have been.

"Love the words...," Burns speculated.

He never thought he would love anything, let alone words or concepts.

But then Burns had another question: "But I had no problem killing the two targets?"

It was easy to see that David was processing a bunch of data and was thoughtful in considering a possible reason.

"From what you mentioned, it sounds like you had no choice. It was either let them kill you or someone else or kill them. Kind of an Ecclesiastic thing … you know, 'a time to plant and a time to pluck...'"

The words jumped into Burns's head as if he were tied into a monitor that was showing him the words directly to his brain: "...Pluck up that which is planted. A time to kill and a time to heal. A time to break down and a time to build up... Ecclesiastes 3:2–2, King James Version."

Burns readjusted his gaze to see that David was clearly astounded by the way his mouth slackened. With bandages covering his eyes, Burns could see David looking at him with his inner eye.

Finally, David spoke, "Well … at least your memories are back. I'm glad some good came out of this."

There was no sarcasm, no guilt or cynicism. Just sadness. The words and tone were just "sad." The feeling was palpable.

I can feel it, Burns realized.

He watched David drift off and could see him jump subjects. Burns saw the pain on the man's face as thoughts of the explosion must have flooded back into David's mind. Burns knew he could never help him the same way David had helped him.

"So do you know what this is all about? The people after you, me, and anyone connected to you?" David asked.

Burns responded immediately.

"Yes."

He was glad to move onto this topic; this was something he could address.

"In my prior life, I did the logistics, planning, and implementation of domestic and foreign counter-terrorism operations. All of my work was top-secret, and most did not comply with the articles of war, military code of conduct, and the Geneva Convention. I operated outside of the law with full knowledge of key officials, and all missions were fully sanctioned by our government but carried out by a private agency. They were also well documented." Burns thought more and then added,

"Prior to my last mission, I obtained more information than I was supposed to get. It was well beyond my security clearance. I shipped out though and was in the middle of an operation when my access was discovered. There was the accident, and then it must have gotten complicated."

"Okay, not to be heartless here, but in all the movies, don't they just kill the person they are worried will talk?" David asked.

This conversation sounds familiar, Burns thought. He decided he would be as brief as possible since he was not sure if David in his drug-induced, grieving condition was ready or needed details.

At least David will let me talk, Burns thought, remembering how Samantha relentlessly pulled information from him. Burns didn't like that at all.

"Typically, yes. But they really wanted to know a couple of things I know before they killed me. With me under constant surveillance and your treatment, they were hoping that maybe they could get the information first. Then kill me."

Burns knew that the way he spoke of conspiracy, murder, treason, and violation of rights must have sounded like he was simply ordering a pizza. Burns went back to the story.

"Problem is ... I don't really have any hard evidence. I have something, but I need to first locate the new site of the operation center, or I have to get more information from the very director of that agency who wants me dead, which is kind of a problem."

"You don't know where your base is?" David was confused. Burns smiled.

Base? That's funny. Like it's a secret base...

Burns narrowed his eyes and stopped smiling as he realized that while it might sound silly, the operations center was, in effect, a secret base of operations hidden in plain view.

David might be right on this.

A smile emerged again over Burns's face as he experienced genuine humor over how at first he thought David's naivety was funny but then he might actually be right in his assessment. Regardless, he was happy David did not see that he was smiling because it might have been construed as patronizing. That was yet another new thing for him - he cared about his companion's feelings and took his time to answer the question fully.

"When an operative like me goes rogue, the operation center's location, servers, codes, everything is shut down and relocated. The data I actually have are the possible locations of where the operation center might relocate to. The new location could be in one of three cities. Problem one is finding it and then figuring out how to either breach its security or get the director to talk or both. Breaching the security might be possible if they haven't found the design flaw I found. Once I get what I need, I might be able to steal enough compromising information to negotiate for our lives or be allowed to live in obscurity without the constant threat of being hunted down and killed. Or another possibility might be to compromise the actual operations center and force them to use their backup operation center, auxiliary control, which may have significantly less security but a short window of operation." Burns was clearly thinking out loud.

Burns noticed that David was listening intently before he posed another question.

"So why didn't they just kill you back at the hospital?" David asked again. Burns looked at him and wondered if he was still foggy or if it was too much data or if Burns wasn't clear enough. He decided to be as clear as possible.

"Before I shipped out for my last mission, I was able to penetrate my agencies' computer server's defenses. I know of a way to sneak something into their computers. I think I've found a design flaw in the majority of the firewalls in my own and other federal agency. Domestic and abroad. In addition to that, I was able to access the code that randomly sets locations for my agency once it is compromised. As a result, I am now able to narrow the locations to three possible sites," Burns concluded.

"Why did you do this?" David asked as he felt his eyes through his bandages.

It's got to be rough for him. His wife and eyesight are gone. His life is now gone, he thought.

Burns tried to answer the question. *Maybe this is distracting him,* he wondered.

"Why did I do it? Partly because it was part of my job to test its defenses. My boss was good at thinking ahead, and he wanted to prepare for any eventualities. Secondly, I like challenges. Anything can be breached. Daniels knew that I found not just one but two areas of weakness before. This was after two years of their reconfiguring their security. Daniels wants to know what I know."

David yawned for a moment, and Burns thought maybe the questions might be over and David wanted to rest.

He's got to be exhausted.

"Can't they just figure it out?" David pressed on.

"Maybe, but the program's error and corresponding flawed software is so elementary, it would be very difficult for them to recreate my search parameters to find it," Burns answered.

"Are you sure your base has already shut down?" David asked innocently.

"Yes. You see," Burns elaborated, "right now, my old team is missing one of their own — me, a high-value asset who has a lot of classified information, including the possible location of its next place of business. The support team that was watching me is presently out of contact with their handlers, and when they find the entire team is dead at various locations, the asset, me, moves from a security leak to a security threat. Now add the missing 'peripheral contacts,' you and your colleague, no bodies, just missing, then we are all threats to national security on par with terrorists. While they can't openly involve a full-blown manhunt with local authorities or other federal branches, they will have every overt and covert operation just under their purview monitoring all federal and local law enforcement channels. They need to find us and make sure we don't talk."

Burns looked back over at David. He knew that was a lot for anyone to think about, let alone an injured man who had witnessed his wife's death. Burns realized that David seemed to have a hard time with the knowledge that Samantha had killed someone too. He couldn't tell which thought bothered

David more — how another person he knew had almost been killed or that the person he had thought he had known could kill. All this on top of watching his wife murdered.

That's quite a lot, Burns thought.

"That support team you mentioned before when I was still woozy ... is that part of your old team?"

"No," Burns answered, "A support team is usually either private security for more 'legal' duties, or we employ professionals who are connected to organized crime or independent contractors who will kill, kidnap, torture, or anything outside of the law. Due to limited resources and the need for plausible deniability, we would typically have a second-level operative work with two individual contractors to ensure clear communication and make sure mission objectives were successfully completed. I killed the contractor who was sent to kill you. Our nurse killed the other one, and I killed the team leader when he was going to kill me."

The air was still after the last description of the procedural operation of killing US citizens.

Much to David's credit, Burns was impressed by how well he was taking in all the information ... *as crazy as it sounds.*

"So what's next?" David asked.

Taking in a deep sigh, he had reservations saying what probably would happen in about an hour's time from now.

"For them? I'm sure Daniels has a death notice on our heads. He's no longer interested in what I know. I'm a liability." Burns stopped there. He really didn't want to say the rest. David said the obvious.

"So are we. Samantha, me ... Jenny—" David's voice

trailed off. He cleared his voice, trying to focus on something else, Burns guessed.

"So what do we do? What's next for us?" David asked.

Burns opened his mouth, but he said nothing. He had no answer. And based on his last communication with his former nurse, there was another woman and a child he had to consider. It all had gotten so complicated so fast. Then a phrase popped into Burns's head.

"A time of war—"

Both men fell silent. Burns was figuring out a plan. As he had explained things to David, he had reviewed things, which in turn gave him some new ideas. He could actually use his knowledge to effectively predict his former employer's next step. While simple, a course of action that could tip the balance became very clear. It would, however, require time, patience, and the assistance of his "new team," which was composed of civilians.

I can't ask them to do this. How can they? They aren't trained for this. Even if they wanted to, how could they? Still...what are the other options?

It was not lost on Burns that he was definitely thinking more outside of the box than he had ever done in the past. Maybe the fact he was working with civilians wasn't too bad. He was trained to keep things professional and never take anything personally. If you enjoyed your work like he used to, that was a bonus. It was easy to get motivated by having a vested interest in the mission. These civilians were now his personal responsibility. He wasn't exactly comfortable with that, but there was nothing he could do with that feeling anymore. His new team could be perfect because it was personal to them and they had a great deal invested in the results. Burns tried to get in his old head to second-guess what his old boss, now new adversary, might be thinking.

Daniels knows I can't go to my contacts or our enemies. He has to figure I will go underground and do it alone. More to ponder, Burns thought.

Becky missed Samantha very much. Whenever she didn't hear from her for a long time, she worried. Worry typically translated to stress.

Just calm yourself. There's nothing you can do but wait. Just sit here and wait. She'll be here. No one can disappear like Samantha, she thought.

Becky waited quietly in the congested, cavernous lobby as she counted backwards by threes. Then another thought popped in her mind.

Did I empty the garbage can upstairs? I'm sure I did, she thought as she slowly stood up. Thinking if she did or not, she replayed how she organized her room, cleaned the bathroom, made the bed, folded the wrappers she wanted to keep and then took the trash out.

"OK. It was the fifth thing I did," she said quietly to herself.

It was the third or fourth thought that had popped in her head since she had been sitting there waiting for her sister to show up.

Before sitting back down, she re-adjusted her jersey to make sure she had easy access to her pockets where she kept four neatly folded, recently ironed handkerchiefs in one, and the used ones in another to balance it out. All the waiting had made her sweat even more. Even when she tried to get some new clothes at the store, she not only had a hard time finding her size, but holding onto the baby as she searched high and low for large elastic sweats made her perspire even more than usual.

Settling in the large chair, she heard some creaking in the back legs just like she did the last four times she got up and then sat right down.

It's getting worse. That creaking is getting louder. It's going to break, I know it.

Not wanting to move again, she kept worrying if the chair was going to give way. After a painful five minutes of thinking about how she would get up from the floor if it did break, and how her sweating was now all over her face, she decided to move to a large couch that was fortunately right next to her. With her third to last handkerchief totally drenched, she simultaneously pocketed it with the others as she used her right hand to push herself up again. Needing to stand up straight, she moved two feet to her right and then carefully lowered herself into the couch.

Damn! It's too low. How am I going to leverage myself up, she thought as she settled deeply into the cushions.

Feeling her heart race as she imagined all the people in the busy lobby were watching her, she tried to distract herself with counting again so as to avoid another panic attack. Counting backwards from three, she dug for another handkerchief as she felt more perspiration on her forehead, brow and neck pool up again.

Crap! Is it hot or just me?

Sighing as she counted, she looked briefly at her watch and wondered if the baby would be up soon.

At least she sleeps well. I wonder if she'll eat when she gets up, she thought.

Looking through the lobby windows, she saw the Chinese restaurant she had seen before when she checked in. She was able to sample their dumplings, rice, chicken lo mein, and wings earlier, and wondered if she should get some for Samantha. Looking to the right of the restaurant, she also noticed a Lebanese store and wondered if they were open. Shaking her head, she shifted her focus back on the Chinese restaurant as she felt the need for more noodles and rice creep into her thoughts.

Feeling her eighth handkerchief reaching the point of saturation, Becky sighed again as she thought of needing to go back upstairs to see if she should get more as she was down to her last one within forty minutes of waiting.

Damn. I knew I should have simply taken the whole bag of them down. Now I can't wait for Sam. Nice job!

Looking at the distance between where she was sitting and the elevators, Becky decided that she would start her move now so she could get back faster. Edging herself to the lip of the couch, she found enough wood and arm to push herself up on the second try. As she stood on her feet, she began to amble when she heard her name.

"Becky? Wait up," she heard Samantha say in her light voice.

"Thank God!"

Turning to face her little sister, she waited for her with her arms extended. As Becky embraced Samantha, a feeling of comfort and safety filled her heart. Samantha let go of the hug first, of course. *She always does.*

The fact that Samantha let her continue hugging her was a testament to how much Samantha actually trusted her. Even in the crowded hotel lobby of the Marriott, she only saw Samantha.

I'm never going to let you go. Just let me hold you...

Once she cleared her eyes and really looked at her sister, it was easy to see something was wrong. The weak smile, downcast, pouting look, with her eyes looking everywhere – it all spoke to Samantha being more vigilant than usual.

"What's wrong?" Becky asked as they walked slowly to the elevator together.

Under normal circumstances, Becky knew that Samantha would say that nothing was wrong or at least tone down the problem and minimize things. The fact that she was readily talking at a high rate of speed with minimal intakes of air confirmed that there were even bigger problems on the horizon.

Holy crap. What shit have you gotten yourself into? What shit are we all into now...

As Samantha relayed the whole story to her, Becky marveled at her sister's abilities to survive on the streets. She knew that her sister, her "Pumpkin," was a prostitute. She also knew that Samantha could get anyone to believe anything she wanted. Men believed she cared only for them. The nursing school board had believed she had wanted to be a nurse. Samantha had told Becky that she had even convinced one of her professors named David Caulfield that she had given up hooking. Over the years, though, this Caulfield guy seemed to make an impression on her sister. Samantha got interested in school and learning. He was a great influence on her, Becky had always thought.

He wasn't a pig either. Someone you can actually trust. And he made her happy and feel good about herself. Not many men can do that.

Becky thought it was great that Samantha had finally met a guy who didn't like her for just sex.

While taking longer than expected, they had crossed the lobby to the elevator, waited, entered, and finally arrived at Becky's hotel door, giving Samantha enough time to give a nonstop, matter-of-fact report.

Stopping for the third time, Becky caught her breath and brushed away the perspiration with her sleeve so she could better attend to what Samantha was saying. For her part, Samantha summarized the whole story from her meeting Burns and getting him help, her leaving the hospital, going back to her other profession, being warned by Caulfield that she might be in danger, the ensuing attack, the murder, and the call from Burns, who was on his way to New York. Worse, he now remembered being a spy and had saved his therapist but not the therapist's wife, and now she had to have a face-to-face meeting with the him so that she could take David, Becky, and the baby and go underground until things blew over.

Feeling the sweat plastering her clothes to her skin, Becky pulled her too-tight jersey over her butt as she tried to make sense of all of it.

Tony's gone and Samantha's in trouble, and now we have to get out of here...can this get even worse, she thought as she started to look for her magnetic key.

Unable to find any response that made sense, Becky finally said what she had wanted to for a long time: "I was hoping to meet this Caulfield guy, but under different circumstances."

Becky didn't intend to make Samantha feel bad.

"I'm sorry, Becky. I really didn't want to make things worse—" Samantha started to say.

Feeling more sweat drip on her sleeves as she tried to make the magnetic strip go the right way, Becky realized how she must have sounded.

"No, Pumpkin. I didn't mean that. I actually meant I did want to meet him. You seemed to like him, and he's actually a nice guy. I'm sorry he's hurt. None of this seems fair," she said.

Becky's heart ached for her brother. His death was so raw and recent that she was still in shock. With Samantha in trouble, she needed to get focused.

She needs me. I need her, she admitted to herself.

Becky was still struggling with the door when she realized Samantha was not the only one that needed her. At last, the door finally opened.

The baby needs me too.

The room was small but nice. Becky became self-conscious of the fast-food wrappers on the table. While there were many of them, they were stacked neatly.

Now why did I want these wrappers? I know I can use them for something but for what? I said it out loud no more than a couple of hours ago so I wouldn't throw these out. But for what?

Over time, her obsessiveness about cleaning, folding, and personal hygiene started to annoy even her, especially when she forgot things. Becky knew all these traits and behaviors were more about control than "just actions" or some "deep-seated meaning." She hated the fact that the folding, cleaning, washing, they all just made her feel good. It merely got worse when she felt her world was out of control, which it was.

I really should get help for this...I guess medication and treatment won't be an option for a while, Becky thought, as the idea of going off-grid was settling in.

"What happened to that?" Samantha asked as she pointed to the hotel room scale that was unceremoniously tossed in the waste basket.

"It broke," Becky said angrily as she tried to move faster across the room before her sister could ask any more questions. Even though she was feeling winded again, she found herself still feeling better that Samantha was with her.

Becky motioned for Samantha to follow and she covered the short distance from the door to the other side of the room with very little effort. Lying peacefully in a crib between a love seat by the window and the queen-sized bed was a two-month-old sleeping baby. Becky looked at Samantha, who looked at the baby before she looked back at her. Becky could see that she was at a complete loss for words.

I know. It's crazy. How the hell are we going to manage this?
Finally, Samantha found something to say.
"Boy?"
"Girl," Becky responded blankly. Becky typically didn't like babies. She didn't know why; she just didn't. Becky had loved her brother, and now he was gone. The only thing he had left behind for her was a responsibility for this little girl who wasn't even his. Tony had loved the baby as if it had been his own, and that was good enough for her.

"How old?" Samantha asked.

Wait a minute...two months or three?

"I think about two months. Not really cute though," Becky added.

"They get cute when they are about six months old," Samantha clarified. Becky looked at Samantha and then remembered: "Ah ... you did an infant/maternity rotation."

"Yup ... and that's about all I know about infants. Does she have a name?"

Becky felt her heart race again as another wave of perspiration began to pour out of her body. Feeling tired, she tried to remember every interaction she had with Tony and his girlfriend that might give her clue as to what the child's name was. She struggled alone with this for days as she realized that she had not been with any adults who would have asked that very question. She was also very embarrassed because Tony had probably told her the baby's name a couple of times, but she either had not cared to remember or just hadn't liked the baby and the situation enough to remember. Becky felt the heat rising on her face.

"I've been calling her Emma," Becky lied.

It was easy to tell that her sister was now looking at her even though Becky kept her eyes on the sleeping baby.

"What do you mean you've been 'calling her Emma?' Is that her name?" Samantha asked.

Frustrated, Becky blurted out, "I don't know her name—okay! I never liked the girlfriend, didn't like the situation, and now Tony's dead!"

The anger that spewed from her mouth surprised even her.

It was easy to see Samantha soften as she reached out to hug her again as she said, "I'm so sorry—"

Becky stood still as she felt her little sister actually try to hug her, protect her for a change. She was sure she would never stop crying as she felt Samantha try to hold her even tighter.

"I like the name … Emma," Becky said weakly through her sobs.

"It's a nice name, Becky. It's really nice," Samantha said sweetly.

Becky felt she was being held up by her sister alone. She felt as if she no longer had strength in her legs to support herself anymore. With her aching knees weakening, she moved as quickly as she could to the recently made bed. Slowly sitting down, she felt more than she saw her sister sit down beside her as she continued to cry. All the while she wondered how nice it was that there was a baby peacefully sleeping feet away, and Samantha taking care of both of them.

Three hours later, Samantha was on the roof of an exposed garage in the middle of New York City, looking for a four-door sedan.

It was easy to spot the car because it was close to 4:00 p.m. and the garage was clearing out. In the far end of the lot, she approached and found one occupant. At first, she was not sure if the person sitting was alive or dead until the person's head moved to the left and right. As she approached the car, she looked at David, who did not see her, even though she had been right on the other side of the glass.

Oh, David. What happened? What shit did I get you into?

She first noticed the bandages covering both eyes and more wide bandages on his neck going down to his chest. She knocked gently on the window, and David jumped.

"Sorry," Samantha said.

David caught his breath and opened the door. Samantha kneeled down to eye level.

"I'm sorry," Samantha said again. She had been saying that a lot these days.

"Thank God you're all right, Samantha" David responded. She was touched to think that in light of all he had lost, he was glad she was all right.

It just isn't fair. He's lost everything. Everything because I asked him for a favor. No...I didn't even ask. I just had Burns transferred to him. He never had a choice. Damn it Sam! Does everything you touch screw up?

As she ruminated, David pulled a cell phone from the car seat and gave it to her.

"Alex is going to call," David continued.

Alex? Samantha thought. *David is on a first-name basis with the guy who is the reason everyone is losing everything?*

The cell phone rang. Samantha looked at it, and so did David, though his was solely out of habit.

"Speak of the devil, and he shall appear," David commented.

"Stay here. I'll be right back," Samantha reassured.

David said nothing. He just sat there. Samantha knew he wasn't going anywhere.

"What!?" Samantha answered with a bad attitude she didn't try to hide.

Fuck people skills, she thought.

Burns was to the point.

"Here's the deal, Ms. Littleton. I need you to hang low for about six months until I find the ones responsible for all this. There are three potential cities I need to search, and then I have to come up with a plan that will work to either get leverage or buy your way out of it."

OK. And 'yes,' I was born last night and will believe everything you say. Do I look that stupid!?

"Come on, Burns," Samantha interrupted. "You really think I expect you to ever return?"

Undeterred, Burns continued as if she hadn't said anything. "There are some clothes that might fit David in the trunk along with forty thousand dollars. I want you to go to a moderate, low-end, extended-stay suite and sit tight. I will call in two weeks with instructions on where to go next. I would get out of New York. Keep this phone safe. It's the only lifeline you have."

Forty thousand dollars? That's a lot of money to just give to someone. Maybe he actually means what he says?

Samantha hesitated and then had to ask the question that was on her mind.

"Do you really think you can get us all out of this?"

She really didn't want to sound too hopeful.

Samantha gauged Burns's honesty by the response, content, and time. He was quiet for the longest ten seconds Samantha had ever experienced. The content was not much better.

"I don't know. I'm just not sure. I know how my organization works, but it will take time. If they really think David is dead and if you really can disappear, and I get really lucky and find them, it might work. Are you taking your sister and the baby with you?"

Samantha's answer was simply yes.

"Okay. There's a revolver that uses .38-caliber bullets. It has good stopping power if you're close to your target. Have you used a weapon before?" Another simple answer for Samantha — no. Then Samantha realized that a "knife" counted as a weapon. She was sure he had meant "gun."

"Then don't use it unless you absolutely have to. The car your sister used, I would get rid of it and take this one. I switched the plates on a car that already left."

It was quiet for a moment, and then Burns added,

"And Ms. Littleton? Don't work. They will assume you need money really fast and be on the lookout in every sex industry venue to find you. You need to stay off the grid for three months at least. Do you have the phone of the person who attacked you?"

"Yes," Samantha answered.

"Leave it behind the car. Any questions?"

"Are you really going to help, or are we totally screwed?" Samantha had to know.

She waited for Burns to answer. His answer, while slow, seemed believable.

"I am really going to help, but we're all pretty screwed. I do owe you both. If you don't hear from me in two weeks starting today, you need to assume I'm dead. You need to note that today is Wednesday, so in two weeks on that Wednesday between 4:00 and 5:00 p.m., I will call."

Burns added something that had Samantha believe that he would call: "As my former therapist would say, 'This is a marathon and not a sprint; success will be measured in small steps, not miles.' But a success is still a success."

Samantha observed her feelings. She was getting better at this chore. She had attended a lecture David had given where he had used those very words.

"Okay," Samantha finally said. She hung up the phone, walked behind the car, and placed the other cell phone in place.

As she entered the driver's seat, David asked her the same question she had asked Burns.

"So how bad is it?"

"Real bad," she said.

"Do you believe him?" David continued.

"Yes," Samantha responded. She was surprised by the answer almost as much as David was. Samantha knew that David was aware of how distrustful her nature was, so if she believed in Burns, either he was really sociopathic, or he was for real. Because Burns had not killed them both, Samantha independently ruled out psychopathology.

As Samantha made her way to pick up Becky and Emma, she was deep in thought. Samantha knew she had to think in terms of years until things got better. She could think in terms of years, she reminded herself. She had fought cancer, completed college and

nursing school. Still, Samantha couldn't understand why she was smiling until she saw Becky waiting for her. Then it became suddenly clear: she was with the two people she knew cared for her without judgment.

Kind of weird to think of that as you're getting ready to hide underground from people who want you dead, she thought.

Chapter 11

"Omnia vincit amor; et nos cedamus amori"
"Love conquers all things; let us too surrender to love,"

– Vergil

Present Day – May 2

TODAY IS MAY 2. Well, it's been a long time coming. Hm... more than four and a half years of off-grid planning with no run-throughs or tests. What other team could do that? Burns thought as he reviewed the schematics of generators in his head.

Burns and his team had waited years to put this together and pull it off. Burns remembered how doubtful he was years ago, recalling how his team was formed more out of desperation than choice. He remembered his first phone call with Samantha on the bus when she had made him feel guilty, compelling her into helping them.

Boy...she really hated me. What a spitfire! If you asked me then, I never would have thought of her as a team player.

He remembered almost fondly his conversation on the

phone when she had come for David and pushed him to tell the truth and to commit to trying to keep them all alive. He was impressed with his new team's abilities of researching, acquisitions, reconnaissance, planning, and improvising.

Well I'm glad you found me, he thought as he took in the cool early morning air and warming rays of the sun.

While he mused, he started to think of the next steps that needed to be taken if everything was to work. It took him only five minutes to find the electrical power junction and separate backup power to the police and fire department in North Reading. Neither was clearly marked for obvious reasons. It was 7:45 a.m., and he expected Samantha at any moment; however, he knew she would have to put the ambulance somewhere it did not draw attention. Prior to leaving the car, he had removed his paramedic jacket, retained the black slacks, and now donned a white shirt and black tie and a SWAT vest. With the addition of the .44 Magnum and the required aviator glasses with proper FBI identification, he would easily pass as some sort of law enforcement or federal agent. Now in the possession of high-yield military explosives and remote detonators, he would have a viable excuse for being at the standby generator, should any utility workers see him and ask questions.

Now what would the odds be of the utility companies coming out right now to service this remote sector? Still, they would want to know why an officer would be out here. I guess the truth could easily work based on what's going on in the area. Why lie when the truth can work? I wished I had some make-up though for these scars…they are too identifiable should anyone really take notice.

Even as he carefully worked at the circuitry and wiring, he was amazed how much his scars were itching. Still, in regards to fitting into the local landscape, he figured he was safe. With the "coordinated terrorist attacks" he was now hearing about on the news and police scanner, he needed to blend in as part of the solution and not the problem. As Burns was finishing the first set of remote explosives on the primary power junction, he heard someone behind him.

Damn it! Are you just getting lazy, tired or distracted, Burns?
Before he could turn, Samantha identified herself.
"A little jumpy, aren't you, Burns?" Samantha said and smiled.
"You know it's been more than three and a half years. You might want to call me Alex,"
Burns continued as he wrapped up the first rigging, tested the connections, and then moved on to the emergency backup generator about thirty feet away.
Samantha followed and began her questions.
"Is the laptop in your car?"
"Yes. Be a lookout and watch out for me so I can focus on this," Burns requested.
"Are you going to ask me nicely?" she asked, standing akimbo sporting a sly smile.
Sighing he tried to ignore her except to say, "David may think I've changed from a sinner to a saint, but I can still kill you."
"Oh, I know you are a sinner, but you have done more than kill me," she said quietly.
Burns...just stop. It looks like it would be an easy response but

just don't do it. You can't win; you never do. She is way too smart and fast for you to even try.

Not falling for the obvious trap she had set, Burns did his best to focus on the second generator and just ignore her. However a burst of heat welled up and broke out on his scalp as sweat emerged on his forehead. He knew his face was red with embarrassment, and he knew she knew. That was another thing he could no longer control; he had difficulty containing such autonomic responses as sweating when he was nervous, smiling and laughing when he was happy, and anger when he was pissed.

Great. The sweating is just going to make my itching worse. She knows this. Why does she always do this? Control. It has to be all about control.

Burns attempted to pretend that he was unaffected. And that was when he turned, and she was smiling.

"Screw you."

That's it? That's all this covert, counterterrorist, and espionage-trained specialist can come up with?

"I think you did a couple of times."

And once again, you lose the fight and the battle of wits. I'll stick with blowing things up. I say that all the time and still fall for her traps, he thought as he felt the corners of his mouth go up.

He never could win against her. Of all the people on the planet, this group of people had wormed their way into his life and given him meaning. David was his friend. Samantha was his lover. Becky his sister, and Emma ... *well, Emma was everyone's baby.*

Burns did find it amusing that it required trauma, brain

injury, and a need for drastic actions to pull everything apart and put it solidly together again. Big risk meant big gain. He trusted these people with his life, and their investment in the mission's success was more than he had ever seen before in any operation. And he was probably the most anxious: *bad things happened to good people.*

As if she was reading his mind, Samantha said, "Strongest steel goes through the hottest flame."

"How do you know when I am worried?" he asked as he finished up and tested the second rigging on the backup generator.

"You get all quiet and reflective," she said simply.

Burns continued with his wiring. He had done this a million times before in his past life. It was like breathing. It was easy for his mind to wander when he did simple tasks as these. Burns smiled as he fondly recalled how things changed for him and Samantha.

Two Years, Three Months Earlier

How they had ended up lovers was surprising. He was sure she hated him in the beginning.

With good reason, he had always thought.

From the very start, Samantha had always been the person to interface with him at critical junctions of the planning phase over the course of four plus years. As she spent more time with him doing surveillance and ultimately sharing motel and hotel rooms, he found himself attracted to her. But it was very different. It wasn't just that she was physically attractive. It was something about her as a person. Burns did something he was sure he had never done in the past: he did not act on his impulses to just have sex with her. Burns did not make any advances. At the time, he was positive that behavior was new. As time went on, they would eat most of their meals together. He picked up on her likes and dislikes in food, such as her fondness for spicy foods and her disdain of rice pudding. He noticed she spent a lot of time putting makeup on, even when they were just going to be in the car. Burns marveled at her endless enjoyment of romantic movies as he desperately tried to sleep at night. Above all though, he could see that she always fought falling asleep. And finally, when she did, she would first sleep soundly and then wake up almost startled and ready to fight.

Then, after fourteen months of staying at a series of low-end motels, just as he felt he was getting closer to his target's new hiding spot, Samantha got sick. It started out as a cold and then turned into the flu. Still, even though he let her sleep and stay back at the motel to rest, she was not improving. He broke his own protocol to get some advice from David as adjunct to his own first-aid knowledge. Ultimately, Burns had to stay with her. When she was burning up with fever, he placed her in cool baths. When she needed assistance in keeping hydrated and eating what little she could, Burns was there. He physically carried her to the bathroom, changed the sheets, and helped change her clothes. This was a major departure from how Burns remembered himself. It was as if he was watching someone. She initially fought him, but she had no strength. He could tell it was hard for her to let someone take care of her back then.

Samantha was really ill and down for seven days. Burns

was surprised that he seemed to not mind that he was not hunting his prey. He tried to convince himself that he needed "his partner" to be healthy and that's why he was so attentive. Again, Burns had no memory of taking care of anyone else but himself.

Did I have anyone in my life to care for? he had often wondered.

The last three nights were her worst. Burns could tell that Samantha's breathing was labored, so he pushed the beds together and had her sleep with her head and chest elevated so he could make sure she was breathing. He also prayed. Samantha seemed to rest more quietly with background noise. The praying seemed to soothe her.

It has to be the sounds of a human voice that calmed her, he was sure.

Finally, she was better and started to recover quickly. Burns felt different about her. He had always liked her attitude and her intelligence, but he found her "sweet." An unusual word for her, he knew, but she was sweet and vulnerable.

Kind of funny...you have to become deathly ill for anyone to see that side...

Once she was fully recovered, he thought things would get back to normal, but they didn't.

For three weeks, Samantha seemed quiet and distant. At the same time, she wore more provocative clothes and pretty revealing lingerie at night. She started to shower with the door open as well. All of this was new. Burns made all attempts to just go on as normal — that is, until he made the "big mistake" of bringing her her favorite breakfast sandwich. She had seemed more distant the night before, so he thought he would get her something special to eat. It was easy to remember her favorite meal - he found her choice of breakfast sandwiches revolting.

There she was, sitting on the edge of the bed, looking down at her egg and cheese bagel with mustard and ketchup, muttering.

"What? Did you say something?" Burns asked. He could barely hear her.

"You got me my favorite sandwich!" she said more clearly and with more force than needed or expected.

Lowering his coffee, he stood looking at her. He must have looked confused pretty much because he really was. Suddenly, he felt his scars starting to itch, mostly the ones near his scalp, but he felt more itching radiating to his arms.

He carefully watched her close the wrapping of her sandwich, sit straight up and seemingly collect her thoughts.

Now what in God's name is all this about? Is she just crazy? Is the stress of all this getting to her?

"Are you gay?" she asked.

What? Why would she ask me if I'm gay? Over a sandwich? And an awful one at that, he thought.

Looking at his feet, he was baffled.

"Um ... no." Burns answered.

He kept it short so he could try to figure out if Samantha was falling apart or worse.

Suddenly, he watched her stand up and run fingers through her hair, obviously distressed and trying to put what was bothering her into words. Finally, she stopped and looked Burns in the eyes. He always marveled at her eyes

— one blue and one light brown. She always seemed to keep them covered with sunglasses.

Except recently, he realized.

"Okay. Look. I don't know what you want. You help me get better. You take care of me. I offer you something in return, and you don't seem interested. Are you gay or just weird?"

Offer me something in return? What? What are you talking about? You're the one who's been acting weird! Wearing nighties, showering with the door open – and don't tell me it's because the fan doesn't work. I tried it and it works fine. What are you talking about...

Burns was silent for a moment. All his years as a logistical specialist and a counterintelligence field agent were called upon to figure out what she was saying.

Finally, it hit him. The provocative clothes, sexy lingerie, leaving the shower door open.

Shit. How stupid are you, Burns? Very! I really suck at this. I'd like to think I'm slow but I'm really just too stupid to figure out the incredibly obvious. No wonder I never did the "girlfriend" thing, he thought.

Burns closed his eyes to find the right words. He felt he needed to get this right if he was going to survive.

I have just one shot at this. My God – catching that bad guy on the Golden Gate Bridge was easier than this. And he was armed with weapons of mass destruction, he thought as he slowed his breathing down.

"Samantha, I'm not gay. I am attracted to you. I helped you because you needed help. You were sick. Just like you helped me when I was at the hospital. You don't have to repay me with sex."

OK...just wait. Be silent.

Burns waited. She looked at him, confused.

OK. Did I screw this up? he wondered.

Burns felt relieved when he saw her face suddenly soften, and she sat back down on the bed as she picked up her sandwich.

Oh, great. I embarrassed her. Why do I care? he thought.

She quietly reopened the wrapper and took a bite of her breakfast sandwich. After a long moment and two more bites, she finally spoke.

"Thank you," she said quietly.

Finally...now where's my croissant?

Burns returned to his cup of coffee and thanked God that he could just take a bite of his own croissant and forget about the whole event when she uttered her next question.

"So ... you're attracted to me?" he heard her ask.

Burns slowly turned. He could feel her eyes on him. With all his scars ablaze and itching, he knew his face was red with blush. He slowly looked at her face. Burns could only describe the smile as sinister, or maybe it was more of a leer. Whatever the smile was, it conveyed more sexuality than he had experienced in years.

Maybe ever...OK. It's time to exit. Keys are in my pocket and I'm already ready to go.

Burns acutely remembered turning away swiftly and saying, "I'll get the car ready," and then he walked out as if the room was on fire.

As he walked briskly out of the room, he heard Samantha call out, "We're not done with this!"

Burns remembered turning the corner and leaning against the wall. He couldn't tell if he was excited or frightened.

Is this lust or more? he thought.

"Get focused, Burns," he said to himself. He started to focus on today's targets and objectives.

The day went on, and Samantha seemed to be herself. Actually, Burns thought she seemed happy.

As the weeks passed, Burns could see a significant change in her behaviors. She asked him questions about his past. That was hard because he could only clearly recall missions and assignments and nothing personal. Rather than watching her movies, she would turn on documentaries about "special operations" and military missions. She asked him questions about it. He was not sure if she was really interested in what he said, but she seemed to be.

Then the day came when he got word from David and Becky that they had refined the search pattern and found a more likely area that needed to be investigated. After a week, they caught a break and spotted some very familiar personnel and what appeared to be a major breakthrough with the operations center. After a short amount of time, Burns confirmed that the lead was solid and the location "probable."

Unbelievable! With just a tweak of the search parameters, David and Becky found it. Three states to choose from. Finally we can take the fight to Daniels!

On his way back to the motel, Burns felt good. He decided to get a bottle of wine and flowers. Not roses or anything serious.

Just a small gift for Samantha for all her help, he reasoned.

Burns found that he didn't seem to like alcohol. He did notice that Samantha liked red wine.

He was feeling almost elated when he could smell

something cooking in his room. It was the smell of cooked chicken. As he opened the door, he was hit with the wonderful odor of a cooked meal as it was being placed on the counter. Samantha turned to see him, and Burns saw her smile.

Wow! Now that's a smell I haven't experienced in a while. Too much eating on the run.

After the momentary distraction, he noticed that her eyes were looking at his hands.

"Flowers? For me?" she asked while he did his best to refocus on her rather than the chicken dinner.

She must have found a place that specializes in gourmet cooking, Burns thought.

By the time he had shifted focus to Samantha, she was inches away from his face. She had the flowers in her hands, and she leaned up to kiss him on the lips. It was a soft kiss, and Burns felt warm and close. As she pulled away, he saw her stop again and look deeply in his eyes. Burns held her so she couldn't move away. She didn't resist. He couldn't remember ever feeling spellbound and connected. She moved closer and kissed him again.

The last thing he remembered from that kiss was his putting the bottle of wine down before he dropped it and his pulling Samantha firmly into him.

I've never felt this before, he thought to himself.

Days together watching for clues and close surveillance would lead into nights of lovemaking that Burns could only describe as powerful. His own feelings were actually clear to him. He now attributed every deviation from his prior behavior to either the brain injury or the treatment or both.

Just to confirm his thoughts, he asked David if and why Samantha might be attracted to him.

David was, as expected, reflective and clinical about it. He pointed out that Samantha had always had a "thing" for him, which had started the first time she had seen him at the hospital being victimized by the staff. She then went out of her way to get treatment for him, which eventually led to David's therapy.

What made Burns different to Samantha, David speculated, was the fact that Burns had changed from a cold, calculating instrument of law enforcement to a vulnerable human being who had started praying. Burns noticed David hesitate, but then he told him what he thought was a turning point for Samantha.

"The day she overheard your noon benedictions praying for 'the sheep in your care and especially Emma,' she was moved. There was an even deeper meaning for her. I think she believed that if you could change, she could too."

Present Day – May 2

All these memories of the past four years flooded Burns' head as he stood over a generator and started putting the rest of his satchel of tools away. When he turned, he noticed that Samantha had suddenly closed the gap between them and kissed him on his forehead.

With both hands gently cupping his cheeks, she said, "I need the laptop now. I have to threaten our government."

Oh, yeah ... today is May 2. My God! How long was I daydreaming? In the middle of a mission no less, he thought.

He smiled and produced a USB flash drive.

"And you will need this," he added.

Samantha had changed yet again, and while she had retained her scarlet red hair, she now had her navy colored jacket, matching her slacks with the town insignia and the bright safety lettering "North Reading Cadet Academy" on the breast plate and back. While she had a tactical belt that could accommodate a gun, the holster was not attached, leaving more room for an additional police scanner, a very powerful two-way radio, flashlights, and a bright yellow scanner that was actually a trigger for an explosive. As they walked out of the woods, he realized that it would have appeared strange if anyone saw two law enforcement officers holding hands. Samantha broke first and took Burns's car, which held official federal government plates and the laptop she needed for the next phase of the operation. Burns resisted the urge to go over the triggering device of the explosives, which was firmly attached the car's gas tank.

As he entered the ambulance's driver's side, Samantha looked back and asked, "The explosives are not activated yet, right?"

Burns was relieved she had asked. He reminded her that she would have to turn the device on once in place, and then she could trigger the explosion with the yellow scanner. She smiled at that point and was about to leave when Burns reminded her of something.

"Did you text your sister?" Burns asked.

"Already done," Samantha said as she took out her phone and showed Burns the text.

"Alpha and Bravo rendezvous complete. Package exchanged.

Alpha is on to prime objective and Bravo to launch point two — out."

Burns smiled, and watched her drive away. Seeing that the time was 8:05 a.m., Burns had to get to the police department's parking lot, assuming that everything would go according to plan.

Like that ever happens. "Grand ideas, wild excitement, grilling details, hampered plans, disappointment, catastrophic failure and punishment of the innocent." After all these years, I can still hear Eric saying that as if it were yesterday, he thought as he returned to his vehicle and pulled out a map. Looking at his watch again, he made a mental calculation of when Becky had to leave to get to her next assignment.

Things are really speeding up now.

Chapter 12

"Abnormis sapiens"
"Wise without instruction,"

– Horace

Present Day – May 2

BECKY WAS NOW SHOWERED and partially changed into another set of running clothes that looked more disheveled than the ones she wore before. Prior to putting on all of her clothes, she pulled out her small, portable weight scale, closed her eyes and carefully stepped on it. Opening her eyes, she saw the scale settle on 153 pounds. Stepping off of it quickly, she waited until it stopped moving before she repeated the ritual. Again, the number remained the same.

Smiling at her continued weight loss progress, she was grateful for her second life and second chance to change everything to make her gains. Putting her travel scale in her overnight bag, she tried to recount how she did. Then she laughed.

Yes...live a secret life, plan to attack a couple of federal agencies, live underground and do shit that the most brazen of criminals wouldn't even consider...yup. That kind of change can be great motivation for a weight loss plan. I wonder if I could market it. That's all you need is to have a death notice on your head and treat every day as if it were your last, for real, she thought as she pulled together her next wardrobe change.

For her next performance, she had to look harried and not even dressed for work; she would have to look like a part-time secretary whose child care had fallen apart just when her boss had needed her the most. She was having second thoughts about the plan.

"I'm not a spy," she said to herself.

At 8:04 a.m., she received another text, and her response was brief: "Message received. Prepping for launch at primary location; ETA is 10:00 a.m. Charlie out."

Becky looked carefully at the text and then tried to distract herself so as to avoid negative thinking. Her run was great. She was appreciative that the office had a shower. She felt exhilarated and almost invincible. The donut she and Emma had eaten added to the sugar rush. Emma was now playing with her makeup set and a remote-controlled toy truck. She was happy.

What more could anyone ask for? Except maybe a life that doesn't involve burn phones, surveillance, high-power rifles, and explosives...Okay. Now back to what I have to do next, she refocused.

Becky started to pace as she went step-by-step over her next move. The plan was to show up at the police station with Emma, plead with the front office to give her boss the medication and file he wanted, have them tell her it was the wrong file, and then beg to use a police computer. That meant she needed to look nonthreatening and mommy-like, which could have simply been over her head.

Now that's something I can do with ease.

Initially, that meant she would have to wear no makeup, mismatched, frumpy clothes and keep her hair long but slightly tangled. At the last moment, it made more sense for her to put on running clothes and have Emma in the jogger. This way, she could run to her destination. Running helped her relax when she was stressed.

Again, all easy, Becky sarcastically thought.

Becky knew that her job was very tricky as she would be with police who were vigilant with all the events occurring around them. In her mind though, she thought David's job was really nerve-wracking — keeping the police at bay with his interview and therefore distracted from efforts just outside their own doorstep.

I don't know how he's doing it, she wondered.

Still, for her, she figured the hard part would be for her to put the flash drive with the virus/worm/Trojan in the back of a desktop CPU or some place that it would not be seen. The not-so-difficult part was to send an e-mail with the critical wording to alert Homeland Security, which in turn would send it to the target destination — the Department of Defense Foreign Intelligence Agency's operation center. The timing here was critical. She knew she needed to be at the police department no later than 10:00 a.m. so she could get her part of the job done.

Becky took a moment to stretch her arms and legs while in a fresh set of running clothes. She then went to her bag and pulled out her anti-anxiety medication. She was not a fan of medication in general. Her running and exercise helped her with a lot of her stress and anxiety. But the medication helped her with her anxious thoughts and the corresponding obsessive behaviors. She was thankful that David had suggested that she just try it.

I shouldn't have waited so long, she thought often.

After she got herself together, she would have to collect her props, get Emma, and make the short drive over to the small mini-mall near the station. She would then have to run with Emma in the jogger again, do her best frantic act, and then run back to the office while leaving the car at the mini-mall for Samantha and David to use later. Having the car in place was critical for "extraction," as Burns had put it. This would be her second run of the day, but it would be the shortest — 1.3 miles that would take her about fifteen minutes at the most.

I won't have time for a shower, she thought.

Becky walked over to look out from the building's window, which had a grand view of the interstate highway right at the junction of the off-ramp exit. With her binoculars, it was easy to see the end of the off-ramp, which led to either the police station where David was being interviewed or the private bank where the "treasure," as Burns had put it, was housed. Becky had a hard time understanding what Burns was saying at times. He had an entire language that was not familiar to her. It was similar to how she didn't understand some legal terms and meanings until she became a paralegal. Regardless, she did respect the fact that Burns didn't "dumb things down" and that he expected her to figure them out. As a result, Becky knew that this intelligence that Burns needed was in a secured room above the bank that had a vault.

At some critical point, she would have to be right where she was standing now to inform Alpha and Bravo teams to either head to the private bank for plan B or more likely enact plan A. Becky was far more nervous and as a result the more conservative one. She thought plan B was the more likely of the two. Burns did know the players though and was betting on his former boss's ego and an impulsive push for glory to win at all cost.

Some guy named Daniels. Sounds like a real prize, she thought.

In any event, she hoped Burns was right. Plan A was more thoroughly planned like a surgery; plan B was more a smash-and-grab.

You know, with your luck – it's going to be the smash-and-grab.

Because she was in a corner office, the rear of the office had a very pleasant view of the sleepy town of North Reading. There was a runners' trail that went by the center of town and cut all the way to the facility parking lot of the team's office. This window had a fixed telescope looking at a designated point in the middle of budding green trees. As she peered through the telescope, Becky could see a lot of movement at the police department where David was being held. With all the activity going on in the area and the adjacent towns, it looked like there were more police personnel leaving the police station than coming and remaining.

Maybe that's a good sign. Fewer cops there might make it easier for David and Samantha.

Becky sighed and looked away, turning her attention back to the windows overlooking the interstate again.

Right below the bank of windows were three laptop computers, which were presently streaming news reports of a series of "serious events" and possibly "coordinated attacks" and "questions of national security." As these various reports were announced by local news carriers, there were now national reports showing a series of images. Becky felt guilty as she saw aerial images of a truck with white smoke coming out of barrels in the middle of a parking lot. Another image showed a smoldering building still being doused with water, while across the street, a hospital was evacuated as the bomb squad team was going in. Yet another image showed the local police giving little information of a local "organized crime gang" shooting involving federal agents and organized crime in North Reading. All the newscasters were drawing conclusions that all these events had two things in common: these events had occurred within fifteen miles of each other in the Merrimack Valley, and there was Muslim or Arabic writing making reference to today's date.

"You know...it's not going to take them long to figure out what that date means," she said.

As these images floated in, Becky pulled herself away from the news reports and went to the desktop computer that was connected to a bank of custom-configured servers. Seeing all the chaos that she and her friends had created was just too much to watch. Becky distracted herself again as she looked at the large computer servers. While not a computer person, it was easy to see that these servers were powerful, to say the least.

While the laptops were also custom-configured with an inordinate amount of RAM, the server provided each laptop with an obscured IP address and their own unique websites. The desktop was only used sparingly and had no one's fingerprints except maybe Burns's. His fingerprints would be connected to unleashing a virulent worm that would not only impair the most recently updated antivirus software of any computers on government sites, federal and national law enforcement and intelligence communities in particular, but it would simultaneously attack all virus protection services for the general public. This would affect anyone who was downloading virus definitions to protect their computers.

It's all part of the plan.

Becky had to admit that Burns's cyber-attack was brilliantly evil. In reality, these virus updates would do the opposite by infecting the consumers' computer by first changing their passwords and locking them out with a password of thirty-eight random characters, and then the computer screen would either freeze or pixelate to black as all the files — documents, e-mails, photos, music, anything saved in a hard drive — would all be erased permanently.

Just a computer nightmare. I guess by using this thing, he's sending his old boss a message. Jesus...Burns really is pissed at this guy. Glad he's on my team. Hate to be on the other.

All this would trigger a heightened state of alert — cyber-attacks in conjunction with the seemingly physical attacks. This would push all federal agencies to implement their emergency protocols, all of which Burns was counting on, especially the one that the Department of Defense would implement – *relocate the operations center to the auxiliary site.*

Early on, Becky had some misgivings about the attack, though. David did too. He asked the question that she was having a hard time articulating. She respected his use of language.

"What if this virus compromises some critical systems like someone's life support or an elevator?" he had asked when Burns had laid out details of the whole plan.

Becky watched as Burns became suddenly quiet, and her sister averted her eyes. Those nonverbal behaviors spoke volumes to her.

Well, looks like you two already talked about this possibility and you're not sure. That's just great, Becky concluded.

It was always easy for Becky to read Samantha. Burns had also become more readable as the years had passed. Becky could tell that they both were close now as each seemed to have an unspoken communication between them and were always on the same page.

Becky noticed this shift after Samantha was sick. Becky felt guilty she couldn't help her sister then. Much to her surprise, Burns called David to ask how he could take care of her.

Pretty nurturing behavior for a spy, Becky thought back then.

After that and several weeks later, Samantha seemed not only protective of Burns but happy. Becky had never seen her that way before. While initially jealous that she had been emotionally replaced by Burns, she was happier that her sister had finally seemed to find someone she cared about and who cared about her.

"I'm pretty sure it will only affect e-mails, digital videos, phones, and other systems related to communication. I don't think it will affect other systems. Critical systems typically are separate from those systems," Burns summarized.

At the time, Becky focused on the phrases Burns seldom used, such as "I'm pretty sure" and "I don't think." These were phrases not in his vernacular. That was Burns's way of saying he really wasn't sure. It was more a hope than a fact.

Becky was surprised that Burns seemed disturbed that he could not be sure that no one would be hurt by their actions. That was a relief to Becky.

At least it's not the plan to deliberately hurt people ... but what if we do? What then? she asked herself.

It looked as if David was going to say something more, but he stopped. Instead, he turned his face towards Becky. That was David's way of making "eye" contact. She knew that they both were still fearful that innocent people could get hurt. She also knew that neither of them had another suggestion that was less risky.

It was now 8:10 a.m., and Becky had nearly two hours before her theatrical debut. She took the time to make sure all the luggage was packed and broken into three sets. As she reviewed everything, she mentally checked off each task. The largest set was for David, Emma, and herself; the next size was for Samantha, and the smallest set for Burns.

Depending on how things went, she, David, and Emma would take half the money, a minimal part of the treasure, and none of the critical documents but all of the laptops. Samantha would take a quarter of the cash, a small part of the treasure, a third of the critical documents and external hard drives.

Becky was thankful that Burns knew his stuff. He had made sure that all parties had rented large storage sites where they had stored different cars, more clothing, much more cash, and passports. Each of these storage sites had been rented for an entire year so that their present vehicles could be safely hidden for some time, allowing the trail to go cold. Becky, David, and Emma would head north to Montreal, where they would find a site Burns had already set up. Samantha would head north to Minnesota. Becky knew that Burns would know all of their locations, but they did not want to know his.

The less I know, the better, she thought.

"Check-ins" would be weekly at a designated time and day. No call from Burns meant they would abort each plan, improvise, and vanish. Burns would take a quarter of the cash as well as the majority of the critical documents, hard and flash drives of the operation. Because Becky and David had Emma, they were not going to take any weapons. Samantha and Burns had planned to be armed. Even though Samantha and Burns had taken less luggage, theirs was the heaviest of the teams' for that reason.

Becky was positive that Samantha and Burns would hook up later. Her mind traveled back to Burns and her sister's unique relationship. *I'm going to miss you, Pumpkin. I must say though I've never seen you happier.* You *really do know how to pick them, though. You couldn't find a doctor? A lawyer, maybe? We could all use one right about now,* Becky thought.

Becky returned to her mental checklist and reviewed her future duties. Her first task would be the detonation of the general power supply. Once she saw the car explode in the police parking lot, she would then trip the explosives attached to the backup generator for all first responders. She had no worries that Burns and Samantha had everything ready to go. Becky had seen Burns's dramatic fireworks across from the hospital on the news. The explosion looked fierce and Becky truly hoped that Burns made sure no one was in the building when it exploded. Her next fear was that her actions might kill someone. Becky began to pace as she ran out of things to do. If she didn't have Emma and her sister, Becky was sure she would never carry out such an outrageous plan. If their lives didn't hang in the balance, she was sure that she would have lost her nerve to follow through on her part.

Four plus years ago, Becky never would have thought she'd be helping with this plan.

You know…I could really use a smoke, a box of chicken or noodles, and an ice-cold rum and cola…Thank God I can feel the anti-anxiety medication kicking in.

Chapter 13

"Qui desiderat pacem, praeparet bellum"
"Let him who wishes for peace prepare for war,"

– *Vegetius*

Present Day – May 2

ANDERSEN WAS ACUTELY AWARE that time was rapidly passing as he took copious notes while David gave his lengthy statement. When one pen ran out of ink, he produced another from his shirt pocket. He always had two. A habit from his old days in Guantanamo.

"So where have you been for more than four and a half years?" Andersen asked. Time was getting short, and while he found his witness interesting, he still had no answers.

As if reading his mind, Coleridge — or rather David — said, "I can answer that, but I will have to try to sum it up. Four and a half years is too long to detail."

Thank God, Andersen thought.

Leaning forward, David began the report, "It was hard for all of us from the start—"

Andersen interrupted, "Who is 'all of us?'"

David summarized how the four of them, Samantha, Becky, Emma, and himself, had been hiding out for nearly four months at an extended-stay motel after his wife's murder.

"With the exception of Emma, everyone's sleeping was bad in quality and short in duration; everyone was eating poorly, if at all, and all of us felt like we were simply waiting to die from an unknown government branch of security. If the government didn't kill us, the boredom or arguing among ourselves would. Even though Burns was living up to his end of the deal by maintaining contact and tracking down the bad guys, he was clear that it was going to take longer than he had expected. This operations center had totally vanished, and three metropolitan cities were pretty big areas to cover."

Andersen noticed that David broke from summarizing to emphasize a critical point. Over the past couple of hours, Andersen had closely monitored his witness, and he realized that David would lean back and cross his legs when he thought back to remember meaningful data. David stayed consistent. He leaned back in his chair, crossed his legs, and folded his arms over his chest. Andersen felt like he was getting close to the bottom of this mystery.

David started the story up again.

"Then one night, it hit me. After four months of thinking only of my lost life, my three children, and my wife's death, I found myself waking up early in the evening … maybe 7:00 p.m., to a crying baby."

Four Years, Two Months Earlier

David stirred from another empty, disturbed sleep, but this time, he had been woken by a crying baby. Emma was in the other room, and he had a vague recollection of either Samantha or Becky saying they were going to get food and something to help break Emma's fever. Reluctantly, he got up and headed toward the source of the crying. He had bumped his knee on a chair and swore. He was not remotely used to being blind. He found that it was easier to simply stay in bed and not get up. If Becky or Samantha were there, he would have simply ignored Emma.

I can't do that, he thought to himself.

He knew he was not too far from the roll-away crib when the crying seemed to stop suddenly. As he approached, he could not tell what was happening, so he called out hello toward the crib. He heard a crinkling of the mattress or diaper, and he had the sense Emma was looking at him. David carefully picked her up and held her close. She was warm and had that "baby smell" he had remembered from holding other babies.

"Well, look at you, little Missy. What's all the fuss about?"

He had never had babies of his own, but the smell was unforgettable. Emma didn't cry but rather sat up in his arms. As she faced him, she began to explore his eyes, nose, and entire face with her hands.

"Well, you are a very curious baby. How did you get so curious? Do you get out of the crib enough? I bet you don't."

His holding her was a novel experience for them both, and she seemed eager to explore this new face — no fear, no judgment, no restrictions, no limits, just determination to figure things out. David found himself rocking side-to-side as Emma continued to explore his face, hair, ears, and neck. As time went on, she began to tire.

"Poor thing. Do you have a fever, or are you just sleepy from all your exploration? Maybe you just like rocking back and forth?" A moment later, she laid her head on his shoulder, and her little hands and arms rested on his chest. Minutes passed, and he heard her breathing continue into the slow, steady rhythm of sleep.

In the silence and darkness, feeling a little life holding onto him, David became both sad and angry.

What happens to you if I die? What happens to you if Becky or Sam don't make it? Do they want you dead, too? A baby? Who the hell do they think they are? What gives them the right to say who lives, and who dies?

As the anger swelled, the emptiness before evaporated.

No. Not anymore. I can't do anything about Jenny. I wish I could. But staying in bed doesn't do anything for anyone. I can't look back. Just forward, he thought as his rocking back and forth changed to pacing in a small circle in a cramped room.

Are we just going to rot here and wait for them to kill us?

As a clear reason formed, he found himself standing still as a thought sank deep into his soul.

Your past life is over. Done. Move forward. Emma needs you now.

David heard the hotel door open and someone try to come inside without making any noise.

"David? Are you all right" Samantha asked.

Still holding the baby, David said in a low, pressured voice, "We're doing this all wrong!"

Becky interjected defensively, "Look! I'm not one of those natural mommies, okay. I'm trying my best—"

David took in what she said, and then it registered. Nodding in recognition of her defensiveness, it was easy to see the similarities of how they both felt.

Just like me. It's not about you, Becky. It's about Emma and her future.

"Becky, I know you're doing your best. I wasn't talking about Emma and you."

It was easy to hear her relief and embarrassment as her insecurities were at the surface.

David heard Becky simply say, "Oh."

"What are you talking about then?" Samantha asked.

Still holding Emma as she nestled more into his chest, he spoke in a low but determined tone.

"We are doing it wrong by letting Alex do all the work while we hide off the grid and wait to be saved. Whomever he is up against, they are our enemy too. They know him, and I bet they think he's alone. But what if we could be more active in helping? Research? Narrow the search? What if we were able to be players in our own destiny?"

David may have been blind but he knew Samantha was skeptical of what he was saying. He could tell she was trying to keep her sarcasm to a minimum.

"David," she began, "we're not soldiers. We're not spies. You're a therapist. I'm a hooker, and Becky's a paralegal. Emma's the only one who is doing her job right."

Come on, Samantha. Can't you see we're more than that? You of all people should understand that — we are more than what others think we are.

"Yes, I know," David said and then tried to curb his enthusiasm. "I know who we are, but are we more than that? Aren't you a nurse? Becky? Whether you like it or not, aren't you a mother now? If we don't become more than we are and get new skills and get in the game, who will take care of Emma?"

David heard silence. He was not sure if he was convincing them or himself more.

Yeah...I didn't think of that either. What happens to her? Does she die too? Do they care? Or will she be lost in foster care? Come on, Samantha! I know you know how that can turn out. You're the lucky one. Will she be as lucky?

David noticed that Becky's breathing seemed heavier.

My eyesight might be gone, but my hearing and sense of smell have improved, David realized suddenly.

Samantha was the first to respond.

"What could we possibly do? What can we bring to the table?"

Yes...you understand.

"That's the best part. Alex is the logistics and analyst specialist. He has created and been in operations. I think there is a body in my grave that's not me. I'm dead. They may not expect me to be of any use. Samantha, you were an unknown to them, and then you disappeared after they failed to kill you. You are still an unknown variable. And Becky ... they don't even know you exist."

"No. I'm not leaving Sam," Becky said adamantly.

While Samantha and Becky argued often, they were clearly siblings. It was easy for David to determine that

Becky was the "mother" in the relationship and Samantha was more the "unruly teenager."

If I didn't live with them, I might have thought it was nice, David constantly thought.

Even though David was constantly sleeping to forget his own pain, he was acutely aware of Becky's loss. It was apparent that Becky was not going to let Samantha out of her sight.

David pressed on with his point.

"Still, you and Emma have the best chance of vanishing entirely. We all have skills, and we have real motivation to do something. They expect Burns to fade away and hide. What if he had a plan and a team of people who are pretty pissed off that their lives are being indiscriminately screwed up because 'we were in the way?'"

Emma stirred a little bit, and David quieted down but continued.

"I've lost everything. I have felt nothing for months now until I felt Emma in my hands, and now I want to protect her. Now I want to get my life started again. I will never be able to get Jenny back. Maybe I will be able to start a new life and see that this one has a shot—"

"Maybe we can make them pay," Samantha chimed in with anger in her voice.

That's right...It's not fair, is it?

"Vengeance is good. If it motivates you to take your life back and keeps us all alive, it will work for me," David added.

David could now feel heat from Samantha. He didn't need to have sight to hear in her voice that she was done waiting indefinitely to fade away.

"Okay, you two superheroes," Becky began. "It sounds good, but do you think our resident spy is going to want our help?"

David was finding that suddenly he had an array of thoughts whirring in his head. For weeks, he had focused on Jenny's death and his own life slipping from him. But now other thoughts were coming through.

Who would take care of Emma if we are all gone?

Ruminating about his wife's murder was horrible, but doing nothing to protect this babe was just as bad, somehow worse. David couldn't do anything to save Jenny. He could only watch her die. He might be able to protect Emma, he reasoned.

I can do something about this, he thought.

Undeterred, David continued.

"I think that once he understands the benefits of more hands, more talent, and more eyes and ears, he will wish he had thought of it himself."

"I don't know, David," Becky trailed off.

David noticed that Samantha picked up his line of reasoning.

"We could use some of the money to set up a low-key operation in Rhode Island and get computers, laptops, and shit. Burns has blank passports that I could get filled with aliases to get us back into the hospital or clinical work with the police or in the courts. We could hide in plain view and use our resources to find them, and then get that leverage Burns talks about so that we can get our lives back."

"How about we get better lives?" Becky added.

Yes! Great idea!

"I like your thinking," David said.

David took a moment to collect his thoughts as his raw feelings of anger seemed to seethe out through his voice.

"Look. I don't know anything about counter-terrorism, surveillance, or spying. I've never held a gun in my life, and the last fight I had was when I was seven ... and I lost. I think that we can help Alex. He may not want it, but anything is better than just waiting here to die, waiting for them to kill Emma."

David knew that saying that Emma might be killed would be powerful. Still, he believed it could happen.

I'm not just going to sit here and watch! Not this time!

David tried to settle his mind down and fell silent. He was still rocking back and forth and holding onto Emma.

This time it was Becky's turn to talk.

"Burns said it could be a year or so to find them. That does give us time to relocate, get money, get resources, and learn how to be — what does he call them? — operatives."

Samantha took the next step.

"All right. I need to meet with Burns face-to-face."

David felt that his mind was calming down until he reviewed Samantha's last statement and wondered if it was truly advisable.

"Are you sure you're the right choice? One of the last conversations you had with him ended with you hoping the bad guys found him and killed him."

Samantha's intake of air informed him that she was shocked.

"You heard that? I was on the phone, trying not to let the other passengers on the bus hear me," Samantha explained.

"Yes. You were loud and pissed. And you were right. It

would have been more humane for him to shoot me when I was sedated."

David actually believed that for months now. He knew he would never actively kill himself, but if he wasted away or Burns killed him, he was sure he would not resist.

The pain and loss would be over, he repeatedly thought every waking moment. That is before he heard Emma crying.

"Sorry. Know I really feel like shit," Samantha finished.

David felt bad that he had embarrassed her but he really had not taken offense to her statements when she had first spoken to Burns. Someone had tried to kill her. He rationalized her words with Burns as a reaction to her anger and confusion, nothing more.

"You know, sis, you have a real way with words when you're pissed," Becky said.

David wasn't sure, but it sounded as if Becky's voice lightened a bit as if she was joking.

"Anyway, I don't have a relationship with him and that may be better in the long run. I can do things you might not be able to do," Becky continued.

"I can take care of myself and Burns," Samantha replied defiantly. David could easily hear the escalation of an argument.

"I can take care of myself too. I can make it work," Becky said, more conciliatory than expected.

It was quiet for a moment until Samantha responded. It was a quiet voice David was not sure he recognized.

"I've done things you haven't, Becky."

David waited for a response from Becky that never came. David could swear Becky's breathing slowed.

David was not sure of the emotion that fell on every word Samantha uttered.

Definitely pain. Fear? Guilt? Shame? he guessed.

Rarely did David ever feel compelled to fill the void of silence. He had learned long ago as a therapist that silence was good.

Not this time, he thought.

"I think we are all going to have the opportunity to do things we normally would never do." David reflected for a moment and then added, "But that was another time and world. Time to change. Speaking of which ... Emma might be a little wet, and the last thing she needs is a rash."

David knew that changing Emma would be anticlimactic. But then this was the first time David had ever changed a baby as a blind man.

Learn something new every day, David pondered.

Burns was surprised by the fresh ideas and a new angle to take.

A team approach would be a complete surprise to my old boss. They expect me to go it alone. Part of my psych-profile, I bet.

He was not sure if his old boss was aware that he had accessed the potential sites for the operations center's emergency relocation, but he was sure that Daniels had no clue that he knew the protocol for relocating critical external hard drives documenting ongoing and new operations at five-year intervals. Burns found himself standing still as he listened to both Samantha and his own thoughts.

But with more hands and eyes working together...With the help of these three, a different strategy with various tactics can be used. *Research and surveillance could be significantly enhanced, and this could literally allow me to be in multiple places at the same time. Coordinated attacks and diversions increase the odds from a limited chance to a possible success. It isn't a big difference, but it could be the difference that separates success from failure. And it would also keep them busy and focused. Hmm...for civilians, they're thinking outside of the box.*

"Look, Burns," Samantha started, "We'll need a list of things you need, what we need to get in order to get established, what we need to do, and how to get ready for whatever the final plan might be. We aren't spies, but David's right - you're the guy that has the know-how to help us get to where we need to be to get the job done."

Burns noticed there was a difference in her voice. He had heard it on the phone when she had broken protocol and called, insisting on meeting him face-to-face.

Determination? Yes. It's all about moving forward and not waiting to die. I get that.

He was surprised she allowed him to pick the place and circumstance. Now in front of him, he was glad he had agreed to the face-to-face so he could see this change with his own eyes. These civilians' motivation and steely determination registered in her eyes.

Burns was still thinking it over, the whole creative proposal, when he looked into Samantha's eyes. She didn't avert them; nor did she try to stare him down.

"What? Is something on me? What is it?" she said as she felt her hair.

It was very easy to see she was more than a prostitute. She obviously had a hard life, gone to school and gotten a nursing degree. Back at the hospital, she saw that he was being kept sedated and somehow got him help.

Yes... it's easy to see that you're more than a hooker...streetwise, you've seen some bad situations and survived. I bet a lot of people miss that. And you've killed before.

Burns looked closely and he was positive she had killed before. Her recovery from killing her attacker was pretty quick, he recalled.

A novice would be immobilized for weeks. She was ready to kill me twelve hours after she killed the female assassin.

Burns had a feeling she was more a kindred spirit than an average "civilian."

She would be very dangerous if she was trained, he thought.

"Do you think David can change? Maybe kill someone? Maybe a bad guy or someone just doing their job?" Burns asked.

Samantha's response was surprisingly immediate.

"He lost everything, and he wants it back. He wants to protect Emma, and I bet he would kill for her. He's pissed, and he's motivated to make someone pay. In the last two days, he's been walking around, outlining the type of computer programs he needs, including text-to-voice ones, public libraries, town halls that might need to be accessed. He's already working with Becky to find free access to blueprints and office buildings. He plans to have a budget for housing, food, exercise equipment, and possible employment that gains access to law enforcement and any agencies helpful to the cause. Yeah ... he's already changed."

"What about you, Samantha?" Burns asked.

"They tried to kill me. I didn't take that personally

because it's a job hazard. But I have my sister and niece with me. I want to kill them all first," she said simply.

Well, that's pretty clear.

"Is Becky in?" Burns asked finally.

"She won't leave me. She doesn't trust you, and if you hurt me, she will kill you. If I kill you, she will help me get rid of your body."

And that's even clearer.

Burns knew there was truth in those statements. Again, he reasoned that her lack of guilt, her street savvy and quick thinking in getting out of being killed in Virginia, all pointed to the fact that she had killed before.

And I bet your sister helped you get rid of that body too. Glad you're on my side.

Then Burns corrected his thinking; *No...Samantha was on her side, not mine.*

"Why does killing me always come up in all of our conversations?" Burns said as he felt a faint smile emerge naturally.

Burns paced throughout their conversation. He had taken to pacing ever since he had left the hospital and ventured out on his own. It helped him think.

I don't know if it ever helped before but it does now.

As he paced, Burns turned suddenly to say something and saw something - Samantha presented a small, soft smile. Burns noticed that her smile included her eyes, and her nose wrinkled a little. He could see that she tried to kill it before he saw, but it was too late. He knew she really didn't want him of all people to see it.

"Samantha, I like your motivation and this whole plan.

But this might take years in getting an operational plan. Can you all hang in there?" Burns questioned.

"Where are we going, Burns? My only time line is that we get this fixed before Emma enters first grade. I want to have Becky and her settled in a nice town with a nice school and neighborhood. That's my only deadline."

"Well, that does give us time to find them and operationalize a plan," Burns summed up as she watched Samantha look through her purse until she took out a small notebook and a pen.

She leaned back against the cement wall of the empty parking garage, poised to take a lot of notes. Burns took a moment and started to talk in bullet-point fashion. As he spoke, he watched Samantha take notes on what they needed to look for in a location, where to buy used materials like furniture and equipment, search patterns to consider, and used computer programs to purchase. Burns especially focused on how to remain off the grid but hide in plain view. A regular schedule of these types of "status operations" meetings was established. A time line and list of exercises to consider for physical conditioning was also worked out. Burns could feel a team was coming together.

Yes. This can work. It will take time but there is a will here.

Present Day – May 2

Andersen stopped his witness in mid-sentence. By the look on David's face, he could tell that his interruption had caught David by surprise.

"You're kidding me," Andersen interrupted.

Now he was worried, and he was sure the intensity and volume of his voice had alerted his witness of his concern. But then again, this was serious — his witness was confessing that there were at least four adults training to become terrorists for years now, with a particular US intelligence agency as their target.

Andersen dropped his pen on the pad of paper. He rubbed his face for a long moment and took a few seconds to regroup before he clearly articulated his question.

"So you're telling me that you, two women, a baby, and a spy were now living together with a plan to get back at the US government?" It was more a statement than a question.

"You know, when you put it like that, it really does sound like we were terrorists," David responded.

What? Are you trying to be funny? You think this is a game? A goddamn game? This isn't a joke!

David's matter-of-fact, clinical response frustrated him.

Andersen stood up this time, walked behind his chair and made sure he was clear just in case he was not hearing things correctly.

"Okay. A group of people living off the grid plan to attack a federal agency. Sounds like Oklahoma City and a bunch of people pissed off at the government to me," Andersen concluded.

For the first time, Andersen noticed that his witness might be more than just a guy at the wrong place at the wrong time. As possible as that was, it was David Caulfield's demeanor that caught his attention – he was completely still. He betrayed no movement until he crossed his legs, folded his hands over his stomach, and leaned back in his chair.

"I think comparing me and my friends to *those terrorists* and their gripes with the IRS, and targeting a children's day care while killing over 150 people is a bit harsh and off base. Anyway ... I have a good reason for my actions. Those people did not." David finally said.

By the tone of David's voice, he could tell that he had insulted his witness. Andersen couldn't believe that anyone would have a good reason to commit domestic terrorism.

"What makes your reason better than theirs?" Andersen pressed.

"The government killed my wife and changed my life forever. They attacked my friend and tried to kill my patient. And when we got in the way, we were considered 'disposable.' They started it," David replied.

You must be shitting me! What are we - eight years old? "They started it?"

Andersen felt his knees getting weak and he had to return to his seat. He was truly exasperated at this man who claimed to be a victim.

"'They started it?' Is that the best you can come up with?" Andersen pushed.

Then David looked to his left and seemed to trail off, and he eventually returned his focus to Andersen and recited a familiar text that Andersen swore he had heard a long time ago.

> *"Two are better than one,*
> *because they have a good return for their labor:*
> *If either of them falls down,*
> *one can help the other up.*

*But pity anyone who falls
and has no one to help them up.
Also, if two lie down together, they will keep warm.
But how can one keep warm alone?
Though one may be overpowered,
two can defend themselves.
A cord of three strands is not quickly broken."*

Andersen looked blankly at David. Maybe his witness was just crazy. Andersen had to consider that as a possibility.

"John's gospel?" asked Andersen. He couldn't think of anything else to say.

"No. It does sound like John though. It is Ecclesiastes, but I can't give you the exact verse and place. Is that better than 'they started it?'" David asked.

Crazy? No. Dangerous? Definitely! He's too organized and too well put together to be insane.

Andersen began to rethink his entire strategy. The closer he got to the personal nature or possible stress points of this witness, the more clinical and resolved his witness became. Andersen was convinced that David was telling him the truth.

It's just that the whole thing is so unbelievable, he thought. Andersen could not afford to have his only key to this bigger problem stop talking, so in the finest tradition of interrogation, he changed his strategy and took a softer approach.

"So how did you live as a group off the grid for more than four years?" Andersen asked.

That question seemed to be more difficult for David than Andersen had anticipated.

So, it's easy to talk about attacking a federal agency but recounting how you lived together as a "family" for four years is difficult. What's up with that?

It took David a minute to regroup, but then he slowly started. "That was very difficult to start, but it somehow all worked," David reflected.

It was apparent to Andersen that he had to begin somewhere.

"At first, the idea of no longer running and hiding was exhilarating, but we were all at odds to figure out how to get started. Burns gave us the first two objectives, which were to set up a base of operations and get in shape."

"'Get in shape?'" Andersen asked.

"Yes," David answered, and then he elaborated.

"I am embarrassed to say that he insisted that while we focused on the logistics of figuring out the day-to-day operations of living together as a team, we needed to physically exercise our bodies to prepare us mentally for learning new tasks."

David paused before he continued, "I know what you're thinking. I bet you're imagining old footage of terrorist cells running around in the desert, shooting off automatic weapons in a training ground. It was far from that. We had to start with the basics."

Andersen had to agree that David was right; he had those images in his head.

It's creepy how he does that.

"Go on," Andersen encouraged as he pushed the feelings aside.

Reluctantly, David continued with his tale.

"The first order of business was to figure out where we were going to live. We decided that we would rent a house, a small house with three separate apartments. This setup would allow us room to set up a living area for all of us, a place for Samantha to work out of, a place to have room to physically exercise, and a place to research and plan.

"In regards to housing, we found the first house right on the Rhode Island/Massachusetts border. Since the house had about three floors, including the refinished basement apartment, we carefully arranged and assigned spaces to accommodate our needs. The top floor was used primarily as sleeping quarters and where we had our meals. This was supposed to be the place to get away from work. We had dinner there, slept there, and rested there. It was our home. The ground floor had a faux bedroom and living area, but this primarily served as our computer and research area. Here is where we would sit and plan, talk and research. This was a place of employment. Here is where a million ideas and plans were born, shot down, and modified. We also set up the exercise room, which was originally a master bedroom, with a treadmill, elliptical, weight sets, floor mats, and a heavy bag. It was a health freak's dream of a workout world."

That explains why you're so fit, Andersen thought.

David continued, "The basement apartment was where Samantha saw clients. She had to shift her clientele more toward sensual and erotic massages. The idea was that she could make about half the money as she was making before without exposing herself. Samantha seemed to be both bored with her job but almost energized to be working toward a greater goal. For her part, she wanted get a life back for herself and to give a much better life to her sister and Emma. She was really changing."

Is Samantha your weak point or is it the little girl, Emma? You softened when you mentioned their names...it's something. One or both of them.

Andersen watched as David took a moment to reflect before he continued.

"The story for our being there was that the Millers, meaning me, Becky, and Emma, lived upstairs. I was a disabled war veteran, and we were married. The second apartment was for a sales representative of renewable fuels who was away most of the time. The basement apartment was for the nice, single woman who had a mental health issue that kept her home.

"Rent was paid typically in cashier's checks or cash. Food was delivered to the door. Rent, food deliveries, and the occasional outing for Emma from time to time was never predictable and hopefully difficult to follow. The windows had reflective glass during the day and were open with lots of sun, while at night, the shades were drawn, and the lights were often kept low. Extra food, clothes, and money were always on hand, and we all had packed bags with the very basics to leave in five minutes. The winter storms we had the last couple of years did nothing to us. We were bored without electricity, but fortunately, we had radios and battery-operated televisions."

Pretty well thought-out for civilians.

"Our mail was delivered to three different post office boxes, and we had two off-site storage sites that could receive packages. Those off-site storage sites also housed two cars that were also packed with more clothes, food or what Burns called 'MREs,' and more money."

Off-site storage, go-bags, meals-ready-to-eat, battery operated televisions...this is a military special operations set up. No, not special operations but more black-ops. These people are the real deal, Andersen thought as he wrote down everything David was saying.

"We set up those debit cards that could be filled with cash and used for Internet purchases for some supplies like medications, vitamins, and often special treats and clothes for Emma. Textbooks, magazines, and seemingly mundane stuff people might order online if they were an outdoors man, hunter, or a hiker — various knives, hiking gear, camping gear, personal protection things — all that stuff would be delivered to either the PO boxes or storage sites."

"Wait a minute," Andersen interrupted. "How did you get a driver's license or proof of residency?"

"Prior to our leaving the extended-stay hotel, Burns had obtained genuine passports that were empty. Between Burns's contact with former covert operatives and Samantha's ties to some people she knew, both were able to obtain different passports which in turn allowed the women and Burns to get different licenses. Once there was a place of residency, the rest was easy — electric, gas bills, hard-line phones, Internet, and cable, all of these were done online," David continued."

No...genuine passports that you forged? How many identities do you people have? My God! You can all come and go as you please if you really have this shit.

"All cell phones and call minutes were purchased with cash in different stores. All the cell phones had to be smart phones to have the ability to keep dates, log data, schedule meetings and designate rally points and codes. The laptops were also paid for in cash. But these were a bit tricky in that we had to find small computer stores that built and configured laptops to very precise specifications. For us, the specifications were to be small, lots of RAM, lots of power without bloatware and lots of memory. The software for all of them included voice-activated and voice-interface capabilities so that I could use them. For every reconfigured laptop we had, we had flash drives and portable, external hard drives and reconfigured laptops in the cars stored off-site."

David took a breath and moved on.

"The hard part was making sure that we got at least one to two hours of exercise six days a week. That was very difficult at first, but after about a year, this part of the adjustment was now a pleasure and we were up to more than an hour every day."

Organized, planned time lines, resourced, physical training – sounds pretty "terrorist" to me, Andersen thought as he tried to write faster.

"The next challenge was to come up with a schedule of breakfast, lunch, and dinner that I could help to make without seeing, and have the meals be not only healthy but good. That meant a very organized arrangement of the food in drawers, refrigerators, and cabinets. This required me to team up nearly all the time with either Samantha or Becky. Of late, though, Emma was my copilot."

This was the only time David seemed to have a visible, clear smile, Andersen noticed and made a careful note.

So...it's her. She's your Achilles heel. Knowing that she draws positive emotion might be important later, Andersen thought.

"In the first year after leaving the hotel, we all had lost a lot of weight, gained a lot of endurance and strength, and were now in a rhythm of operating together. The biggest drawback was trying to keep things fun so that we could have some sort of normal upbringing of Emma. She saw me and Becky as her mother and father."

There it was again — an emotion. There's a pattern. The little girl is David's weak spot, Andersen concluded.

Andersen made another note to that effect.

"And yet she had the attention of three adults all the time and seemed to like doing what grown-ups were doing, whether it be on the laptop, smart phone, or treadmill."

After a moment of silence, Andersen could easily see that David seemed as if he was re-thinking something. Before he could prompt him to go, David started talking again.

What was that about? Something you missed? Something you don't want me to know? What's up with the pause?

"The biggest problem after this was narrowing the areas and search parameters for the operation center Burns was searching for. Frustration and anger were mounting, but then there was an epiphany," David said as he sat back for a moment as if to collect his thoughts before revealing a key piece of information.

OK. This is important. Two quick breaks in the story. Something coming up.

"One night, Becky asked me why we needed surge protectors for the laptops and other sensitive equipment. Once she knew they were needed to protect the electronics from surges and frying their systems, she made a seemingly obvious conclusion. If this operation center was the hub of all foreign intelligence gathering around the world, it would use a whole lot of surge protectors and dampeners for a lot of electronics. Therefore, the electrical output would be over the top, especially if it was located in a civilian building or professional complex, as Burns was positive it would be. I have to tell you that when Becky and I told Burns, his response was, for him, over-the-top. He simply said, 'Brilliant.' Samantha and I came up with the next approach to assist in the search. In order to get into buildings, see blueprints and plans, get images and orders for electronics, and truly gain access to anywhere in any building, we needed to have a cover. We decided to create a limited partnership business that focused on green, renewable energy with us as the energy consultants."

"No way," Andersen said.

This was genius — dangerous but genius. This guy had figured out a way to get his hands on any building permit or plan, access any building, and obtain a look at any manifest he needed. Getting by security would be significantly easier for them now. Jesus H. Christ, Andersen thought to himself.

"It only cost seven hundred dollars online with all of us as officers. Our location was a storefront location that Samantha arranged with one of her clients and a mailing address at one of our PO boxes. It took about three months to become proficient in renewable energy strategies and electrical circuitry, learn a whole lot about architecture and blueprints, and obtain new projects with upcoming and pre-existing commercial offices. He was able to come up with the specific parameters for searches based on the probable size of the operation center, the energy output, the number of staff and vehicle requirements, the city and town locations that would allow such a place to effectively blend in and not draw attention, and the corresponding computer and electronic equipment. Burns felt that he had thoroughly exhausted his search in one city, but with this new approach, he planned to retrace his steps. That left two cities for us to focus on."

David stopped again. Andersen, who had been writing most of the time, looked up and saw that David looked as if he was trying to recount something but had lost the thought. Andersen was initially perplexed. David looked away for a moment. Then Andersen figured it out.

Oh God. That's why you kept stopping...

"You found it, didn't you?" Andersen said.

David hesitated.

He's thinking it through. This could move you from being a victim to an accomplice; a witness to a domestic terrorist, Andersen concluded.

It was eerily quiet for a long moment.

David turned to face Andersen and finally responded.

"Yes. It took about five months to find it. One week for Burns to confirm it. Another two weeks to locate the auxiliary command site and key personnel involved in the operations of both sites."

Well...the mystery of where it is and what role you play is a whole lot clearer.

"So where is it?" Andersen asked finally.

Andersen was sure that David would hesitate when it came to answering this question, but much to his surprise, that was not the case.

"It's in the Shaffer Building in Waltham, Massachusetts, overlooking America's Technological Highway," David answered.

"Route 128. Right next door," Andersen confirmed more for his own benefit than David's.

"Yes," was all David said.

Andersen made a note and then put his pen down.

The fluorescent lights' low buzz was the only sound heard as both men sat quietly.

So...what's next?

Chapter 14

"Recedite, plebes! Gero rem imperialem!"
"Stand aside plebians! I am on imperial business!"

Present Day – May 2

If Samantha learned anything from being in the sex trade business, she knew that being someone else was more about attitude and presentation, and little about sincerity. Without much effort, Samantha knew she could be anyone she needed to be; she could act any role that was required. She could be submissive, innocent, seductive, or dominant. She could remain invisible in a crowded room, or she could hold everyone's attention. She could be whatever she needed to be, on cue. In the past, these skills were her best defense. It kept people at bay.

Let's hope today my acting skills are up to the challenge. So far, they've been pretty good if I do say so myself. Burns was right to have me do this...he better be careful.

As her relationship with Burns grew, she relished his attention. She felt vulnerable yet happy at the same time. She never thought that was possible. She never really believed she could be like "the normal girls" who had boyfriends or husbands who loved them for who they were, not just for sex. Every time she remembered how Burns had taken care of her and hadn't expected sex in return, she still would smile.

I'm not going to lose this life now, Samantha thought.

Fear of losing her newfound sense of living often crept in. It kept her focused on what she had to do … *for Becky, David, and Emma.*

As she approached the Andover town offices, she walked with authority and purpose — *all attitude.* With the laptop carefully encased in its bag, she managed to carry a clipboard tucked under her arm. While still dressed in her police academy uniform, she was no longer wearing the jacket and cap. With the Commonwealth of Massachusetts seal on her starched white shirt and the official manner in which she presented herself, she was able to sail right to the information center of the government building. There, she asked the senior-citizen volunteer attendant where the town manager was and then thanked the woman for the help while nodding to the security guard.

It's all about presentation. You'd think people in charge of these buildings would be more alert and ask more questions. Not that I'm complaining, she thought as she oriented herself while trying to remember the building's layout from memory.

After a moment of simply standing in the stairwell, Samantha went the opposite way she was directed to the basement. Once there she easily found the electrical junction box for telephone and Internet connections.

"Not much of a challenge," she said to herself as there was no one there to question her.

All that practice, all that acting paid off.

Having the building plans and schematics were helpful, to say the least. The absence of internal and external surveillance devices made this public facility the most vulnerable to her next actions. Now it was time for the difficult part. She had to quickly disconnect the main junction of the Internet cable, place a split to reconnect the cable so that the FBI laptop would regain access to the Internet. Electronics was not her thing; Burns had shown her this bridging-and-splicing technique. And doing all of it under pressure with a time limit and fitted gloves didn't make it any easier. Still, it was nice that she wasn't the only one who needed to learn something. Earlier, she was able to return the favor by teaching Burns how to insert an intravenous line without bruising the patient too badly.

Samantha had typically sent a text to her team when she had completed a task. Breaking from this pattern, she sent a text to Becky just prior to sending the worm:

"White bishop to send package to our friends. Time for black bishop to move out. Bravo out."

Samantha imagined all the poor public employees that were in the middle of something online and lost the connection for a few moments. She did feel a little bad about it.

Once the laptop was attached, she logged on to the bureau's official website. Then she took out a flash drive and connected it to the laptop. Two more clicks, and she unleashed a worm right into the FBI's website. While not expecting much fanfare or anything dramatic, the cyber-attack she just unleashed seemed anticlimactic.

"Well..it's official. I can now be charged with attacking a federal agency. That should get me at least ten years in federal prison. I hear Danbury Federal Prison is nice. They just set up a new female, maximum block for terrorists like me," she said jokingly as a way to calm her nerves.

Saying it out loud, however, had the opposite effect – it made her actions more real, and her label as a "terrorist" more accurate.

I'm really in deep now. All the other stuff was peanuts compared to this, she thought as she looked at her makeshift cyber-attack weapon.

As she completed the final stages of her task, she could have shut the laptop and taken it all with her, but that was not the plan. She left the laptop in place, still attached to the system. Extracting a very dangerous looking block of high-grade explosives with a digital timing device, she carefully taped it to the open laptop. Standing back to look at her handiwork, she was very pleased with the overall menacing image.

"Well, if I saw this I'd run away as fast as I could."

Samantha remembered that when she had stolen the FBI laptop from the pair of federal agents in that coffee shop, she had started a cascade of an orchestrated attack. Somehow, by actually sending a worm right to the FBI and leaving the laptop open for them to easily find, this act seemed more deliberate.

Pulling her uniform together, and making sure she looked innocent, she sighed and thought of what she needed to do next before collecting her props.

At the moment, though, Samantha felt really bad for anyone on a computer in the building. If what Burns said was true, in about two minutes, their software would betray its operating system, and all files, pictures, and official and personal, would be copied, compiled into e-mails, and indiscriminately sent out into cyber-space. After that, all those files would simply disappear, permanently erased, from their hard drives. And to add insult to injury, passwords would change to a random code, effectively locking them out, leaving the operator to watch the carnage on a frozen screen.

Man...that's just cold. Where did Burns find this monster?

Shaking out of her thoughts, Samantha regrouped, collected her empty bag and clipboard, and exited the basement while locking the door behind her.

Once she was in the hallway, she began hearing raised voices yelling expletives. As she exited the building, she could hear the escalating voices from the attached middle school, which apparently shared the same computer servers. Echoing from the windows and out the doors, Samantha could hear the chorus of children and adults crying out some variation of the same thing: "What the hell?! I've lost everything!"

But she kept marching. She was committed.

Thinking of Emma, Becky, David and Burns being hunted all these years and hiding, made her feel angry.

"So..." Samantha said to herself as her fear changed to justification in this overt act of war, "...this is what a terrorist feels like."

Jesus...and this worm is about half as bad as the one that Becky's going to send later to the operations center. Shit.

As Samantha looked back to the town buildings, she pulled at her fitted gloves and entered her waiting car. Her next stop would be near the junction of the interstate and off-ramp to await Becky's signal.

The FBI's regional office in the Boston had been very busy all morning when the attack came. Deputy Director John Helms was in the bureau's control room, watching a number of events occurring just outside his proverbial doorstep. He was convinced that he was watching a series of related events unfold on the floor-to-ceiling monitor that was segmented onto smaller monitors for individual analysts to review. These monitors showed all the images and the writings in Arabic, and it also displayed all the possible leads that his analysts were pursuing.

Maybe someone will see a pattern. Maybe a lead.

Helms, a twenty-year veteran of the FBI, knew enough to know when attacks were coordinated or not.

Random? All occurring within the hour and highly visible. These are all definitely connected...but by who?

He was a big man who could still do the majority of his US Marine drills, and he kept his hair short and his hands ready. Even though he was in great shape for a fifty-five-year-old former marine, he still needed medication for high blood pressure and high cholesterol. He would often have to reassure his wife that if two wars had not killed him, the stress of his job would not either.

It's a good thing I remembered my medication and did an early run. There's no way I can get a cup of coffee let alone go for a run. None of us are leaving for a while, he thought as his eyes darted from one segmented screen to the next.

Helms stood in the middle of his control room, looking at the monitors and listening to the low din of his analysts working. He liked his team very much, but many of them were not experienced; even fewer had a military background.

These young people were smart, but they were much more comfortable with tablets and smartphones than hand-to-hand combat and semiautomatic weapons.

Hm...I guess when flying a key board becomes the weapon of choice in the field, these kids will be well prepared, Semper Fi.

Helms shifted his thinking to take stock of what he had accomplished so far today. He had been on the phone with all local, state, and federal agencies, and was now coordinating all deployment efforts to find out what was going on. He had his people take the lead with the North Reading Police on the FBI shootings while he had the state police, bomb units, hazmat units, and all local police focus on the attacks at the hospital, the building across the street, and a smoking truck that appeared to be filled with explosives and a possible chemical agent. He had spent the past hour with the state governor to release the National Guard to all possible state, local, and federal targets, and he even had police academy cadets out in force to assist with traffic and civilian evacuations in the Merrimack Valley. His boss in DC, who was more of a politician than field agent, wanted more proof that this was indeed an attack.

How the hell could anybody think this is random? How much proof do you need?

It took an hour for his communication teams to locate the signals and transmissions of "remarkably disturbing text messages indicative of either a foreign or domestic terrorist attack." When the text messages were all put together, including the one that had come in a few minutes ago, it ran like a special forces insertion team in the middle of black-op:

"Black knight in place. White knight on the move. Alpha out."

"Alpha, Charlie. White bishop on the move. Bravo out."
"White bishop secured transport. Black bishop, you are a go. Bravo out."

"Black bishop has delivered the package and is on the move. Heading to lair. Charlie out."

"White bishop to send package to our friends. Time for black bishop to move out. Bravo out."

"Alpha and Bravo rendezvous complete. Package exchanged. Alpha is on to prime objective, and Bravo to launch point two—out."

"Message received. Prepping for launch at primary location. ETA is 10:00 a.m. Charlie out."

The corresponding events that went with the chatter made it clear that there was a connection. His boss agreed, and pushed for the chain of command to scramble all armed federal agencies, and for the president to put all US Armed Forces, domestic and abroad, on a heightened alert. The National Emergency Alert System might need to be used to alert the citizens.

Now how do you alert people without causing panic? Helms thought.

He banished the question so that he could focus on his task of containing the crisis to the Northeast of the

United States or preferably to the three communities in the Merrimack Valley. But Helms was beginning to feel hopeful that he might be able to actually get ahead of future attacks and possibly capture the ones responsible. Now that they had the frequency and general location of the cell phones, a new transmission could be discovered as it was being sent.

"We do this right and if the text is long enough, we will be able to zero in on the terrorists' location," Rachael Janeson told him as she coordinated efforts with her two-man team, Gilmore and Johnson.

"I hate terrorists," Helms muttered.

With nearly all of his resources and staff deployed either in the field or crammed behind computer screens, he thought the analyst shouting for him would give him good news.

Helms approached the young analyst.

They are all so young, he thought.

Helms's steps were quick, and he spoke curtly as he approached the analyst. By the time he got to his bank of monitors, he remembered his name was Gilmore, and that he might be the second oldest analyst there.

Maybe he's thirty.

"Give me good news, Gilmore," Helms demanded.

"Director, I have a text coming through now. I was able to get a location on it. It came from a government building or near it."

Gilmore's eyes never left his three monitors. One monitor had the actual text with time and date code:

"White bishop to send package to our friends. Time for black bishop to move out. Bravo out."

On one monitor, there was an actual aerial image of the location, and the other had a map with all the buildings located in the area.

"Sir!" Gilmore said.

Like Gilmore, Helms was alarmed when he saw that the signal was right outside of a middle school.

"Nope! We're not having any civilians, especially a school of children, show up on this evening's news! Children engulfed in flames or chemicals? Not on my watch," he said loudly for all to hear.

Helms bellowed the following orders: "Peters, Thompson, Davenport, Jakes, reroute local PDs to evacuate the school and buildings within a five-block radius of that last signal. Call the school and building now and get them out of there. If there are any bomb, hazmat, or SWAT units left, redirect them there. Kelly, get—"

Helms never finished his last thought as his attention was suddenly drawn to a desperate cry of another analyst ten feet behind Gilmore. Johnson was his name, Helms recalled.

Maybe he's thirty, too.

"Sir! I got something coming over one of our field laptops! It's been logged on for the last five minutes, and it has a real bad worm that's hacking through all our firewalls like knife through butter," Johnson said while surfing three monitors at the same time as Helms approached.

"Knife through butter?" What the hell?

"Get moving, everyone," Helms reiterated so that they would start his prior orders before a list of potential tragedies became tonight's news.

Johnson picked right back up as Helms stood over him.

"This looks like some kind of Trojan worm. It wasn't picked up by any of our firewalls, security, antispyware, or antibotware or anything. Somehow, it just slipped through and sat there for a few minutes. It looks like it's a combination of Conflicker and Nimbda worms. I bet you this is either North Korean or Chinese."

Again, there were more interruptions. A voice from five rows away yelled out, "I can't get into my computer. It's not recognizing my password."

"Mine too," another voice said. "It just asked me for my password and denied me."

"Sir!" This voice was from Janeson. Helms knew she was the oldest agent there, thirty-four with actual military experience and absolutely no people skills. While strikingly attractive with incredible deep blue eyes, she was socially awkward. She was, however, remarkably smart, and typically all questions were answered by her.

I hope you got something today. We need it.

"My computer froze, and I am now watching my folders on my desktop disappear. I recommend we shut everything down; desktops, laptops, servers, everything," she said with urgency but devoid of emotion.

He knew Janeson would have an answer.

"I'm losing my entire hard drive. No, wait ... my external hard drive is deleting all my files too," another voice said.

Gilmore turned to look at Helms, nodded quickly, and said, "We need to unplug everything now."

Helms looked at Janeson, who was not waiting for him to give the order. She was already on her knees under her desk, pulling plugs out to disconnect her computer and peripherals. Johnson followed suit.

"Everybody! Unplug everything! Crepes and Martin, get out to the exterior offices and spread the word. Davenport and Thompson, get down to the servers and shut them all down!"

Helms did a double take of Martin.

Why is an agency accountant in my control room? he thought.

Helms returned to his next steps.

As he was about to give the next set of orders, he became distracted by what he first saw on one of the smaller monitors. The screen was changing from sharp images to fuzzy white. Helms then shifted focus to watch the floor-to-ceiling monitor's images start to pixelate, freeze, and then slowly fade away. That was a nerve-wracking sight. He had seen the big monitor switch off before for maintenance. It would typically turn off in sections with crisp snaps. This was very different. Similar to the smaller monitors, the big monitor and anything with a monitor or screen tied into the FBI Intranet, began to freeze, pixelate, and then slowly fade away. No clean snap. Just a slow fade to darkness with an evaporating afterimage and then a blank screen. It took maybe sixty seconds for every screen to eventually go dark.

Then there was silence. No computer fans, no air conditioning, no buzz from the overhead lights; just the sound of people breathing and futile attempts at pressing keys on their keyboards.

Everything was quiet for at least a minute. With hands on his hips, he surveyed the area as he took in the severity of the situation. He looked at his people's young faces as they gave an "end of days" expression while continuing to aimlessly push buttons on inoperative equipment. Helms likened the expression to a marine caught in a firefight with defective ammunition — pulling the trigger but with no discharge. The only one not absently pushing buttons was Janeson. By now she had taken out a pen and a legal-sized pad of paper with the obvious plan of figuring something out.

Well, at least Janeson is working on the problem, as always.

"OK. We'll do it old school," Helms said to himself.

Helms stood tall, his hands now across his chest, and issued next steps in the darkened, silent cavern.

"All right, people! Get out of the control room and find any space, any laptop not connected to our servers and get those kids out of that school. I want that school evacuated and locked down!"

Helms started herding all of his team out of the "more technologically advanced cave" that man had made. Helms hoped the offices still had paper directories and telephone hard-lines in the building. He also wondered if his control room was the only one that had fallen to this attack.

Janeson was the last one out. She wanted his attention, so she waited to be last. He was impressed with her newly acquired skill of not just blurting out what was on her mind but instead looking for social cues to speak.

Helms looked down at the pad of paper she had been feverishly writing on.

"Sir," Janeson started, "the Arabic writing seen at the hospital, truck, and other crime scenes translates to 'I remember everything that happened on May 2.'"

Helms looked at her, hoping she knew what it meant. Janeson was the smartest on the team and the best at these kinds of puzzles. In a rare moment of accurately reading nonverbal communication, Janeson shook her head, indicating that she didn't know the date's meaning and that she was baffled. He was too, but he could tell Janeson would not let this task go unchallenged. She started to walk ahead of him, clearly making a list of important events that had occurred on May 2 throughout history.

"Go get 'em, kid," he said with a small smirk. It was as close to a smile that he could muster as a result of recent events.

Once outside of the control room, Helms started walking from cubicle to cubicle, getting an update on what was happening. He was relieved that his team had located and rerouted resources to the school. School staff had been notified and had evacuated the buildings.

Helms had to think. His Boston office was without computers, and for all intents and purposes, it was dead. With the control room and its resources nullified, he had to do something. Then it hit him.

In a set of cubicles, Helms stood erect and barreled out orders.

"I need five people to stay in the office and man the phones. Everyone else, we're going mobile. Get the cars ready and bring your weapons and tactical gear."

"Sir," Gilmore said, "where are we going?"

"We are going to Andover to secure the crime scene, and then we are going to catch the bastards who are doing this," Helms said casually.

"Fucking A," said Crepes, who had returned with Martin from the room with the servers.

As his team began to reorganize, Helms walked to his

office alone. Once there, he put on his tactical vest, semiautomatic sidearm, and the standard FBI windbreaker. He was ready to hit the field.

At least that's doing something.

Taking a moment to collect his thoughts before he left, he found his cell phone, with the plan to bring his boss up to speed about how anything electronic and computer-driven, even the digital office phones and cable television in the offices, were now inoperable. Getting voice mail, he paused just to summarize the situation:

"A major cyber-attack hit us and took about a minute to nuke everything once it had been discovered. The good news here is that you can give the rest of the intelligence and defense community a heads-up. Bad news, we're totally dark and going into the field to find a needle in a haystack."

Miserable day, he thought as he closed the line.

As Helms started to walk to get his troops moving, he remained perplexed by the May 2 date. He was close though. There was something about that date that did have meaning.

And it's personal for all of us...I can feel it.

May Day was typically a day of celebrating spring in the United States, Canada, and parts of Western Europe. Cinco de Mayo was May 5, and that too celebrated May, particularly in Mexico but also in the United States. His attention came back when his boss called to let him know that all planes had been grounded as a result of the attack they had just endured. After a few minutes of driving in the car, it became clear.

"Jesus, Mary and Joseph!" he uttered.

Helms knew that his team would not be surprised by the

expletives he uttered. He did that a lot. More swears spilled out in rapid succession as Helms fumbled for his cell phone.

It was easy to see that his team was all being quiet for fear of their boss losing his train of thought. Janeson didn't seem to even notice as she kept writing lists.

Helms finally had his boss's secretary on the phone, and he was really getting pissed.

"Jesus! He just called me. Is he at the Pentagon or the situation room or the bathroom?" he demanded to know.

After more nonsense, Helms couldn't take it anymore and finally announced his epiphany:

"Tell the director that May 2 is the day Oman Sharif Sudani was killed. May 2, 2011."

Janeson, who had been squished in the backseat, looked up and yelled out, "Shit!"

It was clear that she was annoyed that she had missed something so obvious. And for her to have an emotional outburst was unusual, to say the least.

Oman Sharif Sudani was one of the key architects of domestic and foreign terrorist attacks. He was a walking "command and control center," and notorious for being the best in executing attacks. All of his operations were the most coordinated, organized, and logistically planned. Today's events looked a lot like his work.

But he's dead. Is this one of his former cells? An "anniversary" attack?" he wondered.

The day he was killed, the world became a safer place. His terrorist hierarchy fell apart, and the free world was able to rapidly disassemble his terrorist network piece by piece. Still, Helms was surprised how rapidly Sudani's network fell apart.

Helms smiled — the first smile all day.

He looked back at Janeson, who was now looking out the window. *Is she actually pouting?*

"Throw me a bone here, Janeson. It's the only thing I got today."

Janeson smiled as well. That was another surprise because Janeson never smiled.

Quite a range of emotions she's demonstrating today, Helms thought.

Becky was back from her second run that day. Everything had gone better than expected: the car was in place for the extraction, Emma was in the jogger, and she made great time to the police department. She looked disheveled, heated, and flustered, exactly the way it was scripted. It got easier when she told the officer at the desk that she was there for Mr. Coleridge and he wanted some information that might be helpful. The guy buzzed her in and told her to wait by the office. Asking if she could borrow a computer to get the information from her e-mail, he offered no resistance and said, "Sure," after which he brought her to an empty desk.

As a result of the crisis, nearly all first responders were out in the field, leaving nearly all the desks vacant. With a choice of computers, she found one with a large CPU and got on the Internet where she opened an e-mail address.

At the beginning of this phase of the operation, Becky was sure she would be surrounded by police or someone paying close attention to every move she made. Instead, the officer started talking to another officer and then returned to the front desk. There were a number of other civilians calling the stations and asking what was going on in their town and the neighboring cities. As the officer at the front desk became engrossed in another call, and the two remaining officers quickly got their gear on and left, Becky was left alone with Emma, who at that very moment was giving her best rendition of some classic rock 'n' roll song.

Something about a hotel, checking in but not leaving...what, she thought as she tried to focus on her tasks.

While she loved the fact that David loved Emma, and Emma just adored him, his insistence that she learn these classic songs could come up at the worst time. As Emma sang, Becky contained her urge to tell her to be quiet so that she wouldn't draw any unwanted attention.

By design, Becky dropped her keys under the desk near the CPU tower where she was able to quickly plug in a flash drive without incident. Once done, Becky patiently waited for the computer to ask if she wanted to recognize the new software and if she wanted to run the new program. Saying "yes" to both, she now endured the longest two minutes she ever experienced as she patiently watched the rotating circle spin endlessly.

What the hell is on this thing that it's taking so long?

If it hadn't been for her anti-anxiety medication flowing through her brain and bloodstream, Becky was sure she would be jumping out of her skin.

Failure is not an option. If I don't do this, we all either get killed or go to jail. What will happen to Emma? Foster care? Different homes...just like Samantha. No, it has to work. I can't fail, she thought.

Finally, the moment came: "New software successfully

installed." Becky went to the programs menu to un-highlight the new program named *Albatross*. Once completed, she had the urge to make sure the new software and program were actually working. This time, she "accidentally" dropped her sunglasses so she could look over to the back of the CPU to reassure herself that the drive was working.

"Some would call it 'obsessive-compulsive' while others would call it 'thoroughness,'" she said quietly to herself as she clearly saw that the unobtrusive red glow of the flash drive was indicating it was working.

Relieved, she got up leaving the computer on, and exited the large, vacant squad room. She told the officer at the front desk that she was leaving because she was unable to access her boss's e-mail and that she would be back with hard copies.

Once outside, Becky started breathing again and began a slow start of her run as soon as she could without looking as if she was running from a crime scene. She only started to relax when she was within the confines of the team's office. Putting Emma in the smaller room, Becky looked out the windows, while taking off her hat and oversized tee shirt. It was then that she saw a convoy of black and blue SUVs and sedans that clearly looked like law enforcement.

Shit! They look pretty official and in a hurry.

She held her breath again as she took out the binoculars to see if they were heading to her exit. Fortunately, they drove right by it. Becky immediately got on the two-way radio.

"Guys … a whole bunch of federal types just blew by the exit."

"Which town? Andover?" Burns was the first to respond.

"Yes," Becky said while still watching through her binoculars and trying to count how many vehicles there were in total.

Then she heard him ask, "Are you on the road, Scarlet?"

"Scarlet?" Who the hell is that? Oh...wait a minute...

Becky suddenly remembered that her sister, "Queen of Wigs," was now sporting a red one.

How can she wear those things all the time? It's got to get hot! You couldn't pay me enough to wear one of those. Better chance getting me in a bikini!

"I'm gone and heading to our rendezvous," Samantha responded.

Becky waited. Burns was quiet.

Probably because he's thinking....

He did not say his usual "okay," which would indicate the plan was to move ahead with no deviations.

Maybe it's the large number of government cars that were en route. Shit...we're going for a "smash and grab." I hate that plan.

"Were there more than five transports?" Burns finally asked.

Becky had to remember, but then she readily recalled.

"There were at least seven," she replied.

Espionage is really not my best talent.

There was more silence. To Becky's relief, Burns finally spoke. For him, he was actually chatty.

"All is well. Change of plans. Scarlet, go pick up the package now. Tiny, break down the laptops and then push the button. I'm going to the vault. See you all on the other side."

OK...that's it.

Becky's legs felt weak for a moment. She knew this sudden fatigue was not the result of two runs or her medication. It was the anticipation and stress of years of work coming to a head. Becky knew that everyone she cared for would either be safe with the stolen leverage or dead within the hour. As she felt pressure build in her chest, she had an overwhelming urge to find her portable weight scale to see if she lost more weight from her recent run and her heightened anxiety.

No time for that, Becky. Checking your weight has nothing to do with your fears. It makes you feel better but you got other shit to focus on, she reminded herself.

"Think positively and follow the drill. Remember that it's all part of the plan."

Becky remembered that Burns had always said that the success or failure of an operation was in knowing when to stay the course and when to alter the plan based on new data. The heavy response as measured by the large number of cars reacting to the last text and attack must have altered the plan.

That meant that the surgically implemented plan was tossed out and the messy, though direct plan was in motion.

"Nothing I can do about that. Doesn't matter what I want. Just focus, except my part of the plan. That does matter," she said to herself.

It took Becky five minutes to shut the laptops down. All the newscasts stopped as she methodically shut down each laptop. The last news report she heard said that all air travel in and out of Boston had now been grounded. With that ringing in her ears, she went to the desktop and hit Enter, and a program that had been ready for launch for five and a half hours was now running.

Becky knew that the flash drive she had installed in the police computer was now on and sending a message that, if traced, would show that it was originating from the police department, and transmitting to a homeland security office. From what Burns had told her, this agency would isolate the key words and immediately forward an important message to the Department of Defense's operation center:

"I remember everything on May 2, 2011. Operations center, foreign intelligence agency, all present foreign operations are compromised. Cyber-attack imminent."

Attached to that message was a virulent computer virus/worm from hell. Becky experienced some anxiety that bordered on panic. She had just started a war with a federal agency as well as the United States. The attack she had unleashed would start infecting various virus protectors to conceal and distribute a terrible worm. Burns was meticulous in explaining that this attack was more powerful than the one they would perpetrate against the bureau. This program would lock out people from their computers and freeze them only so that they could watch their files disappear unless they somehow completely shut down their power. That wasn't the worse part: it would happen in seconds, not minutes.

Because Becky had once been a paralegal, she knew that her actions qualified as an act of terror. Moving mechanically through her assignments, she kept having the same recurring thoughts run through her head when she thought about being a terrorist.

All I want is to get my life back. All I want is for Emma and Samantha to live again. I want David to be a part of my life with Emma too. Is that crazy?

Chapter 15

"Vulpem pilum mutat, non mores"
"A fox may change its hair, not its tricks"

Present Day – May 2

THE DEPARTMENT OF DEFENSE'S operations center was one of the more unusual agencies under the federal government's auspices. It was a privately held corporation that had a significant amount of federal funding and influence with minimal oversight by the very branch that it shared its name with. The operations center was once completely operated by the Department of Defense, which pulled in the "best and the brightest" from all the branches of the armed forces. There were three purposes of the operations center: communication, surveillance, and data analysis. During the late 1990s, the operations center was the only agency that was doing its own data collection and reviewing all the other data collected by her sister agencies, the CIA, NSA, British Intelligence, and FBI. While the other intelligence agencies did not share or review other sources of data and possible connections, the operations center was quietly ahead in the

counterintelligence and anti-terrorist game. The operations center was the only agency that predicted a series of terrorist attacks, both domestic and abroad, as well as the fall of the Soviet Union and China's political and economic seismic shift.

If the other intelligence agencies and armed forces had listened to the operations center, 9/11 for example might have been completely stopped. In fact, 9/11 was the last stage of a three-prong terrorist attack that was to include similar attacks on California's Golden Gate Bridge and Colorado's Hoover Dam. The operations center stopped both of those but not the one launched from Boston. This fact was kept from the American public as it was believed it would cause hysteria and fear. The US government didn't want to give too much credit to the terrorists. In addition to preventing these attacks, the operations center prevented an attack at England's Heathrow Airport and France's Eiffel Tower.

As a result of early intelligence success, there was a need to put some distance between the federal government and the operations center so that it could work more independently without being restricted by the rules of engagement. The center was becoming more effective in finding its targets and "resolving" the issues quickly as well. That meant it could eliminate or neutralize threats faster. Sometimes the threats were foreigners, sometimes allies, and sometimes civilians. To do this without compromising national security and to allow for plausible deniability, the agency became the only federal agency to become private with its own board, private stocks, and governance while at the same time still maintaining nearly unlimited access to federal funding and resources. In order for the agency to work well, it had to be kept secret, and if possible, it needed to hide in plain view. Naturally, most of the analysts, operators, and field operatives were ex-military or military on loan from the armed forces. As time went on, the operations center became the firewall against any nation or organization that threatened US interests, domestic and abroad. Without federal or public oversight, the operations center dictated its own methods, objectives, and missions. Only a few key people were aware of the nature and breadth of what the agency was doing.

Director Thomas "Steel" Webber was the man who had many of the answers. He knew where the bodies were, both figuratively and literally. Still active in the field, he prided himself on finding younger talents who were very effective in their jobs. His personal preferences included attractive women who were smart and lethal. Though he did like men personally, he preferred them in his private time than on his senior management team. He was not a naturally handsome man by any reasonable standards; he was short, balding, and overweight. However, he saw himself as a powerful mover and shaker. He was "the King Maker" of young women's careers. From his home office in Virginia, he was reviewing two managers who were diametrically opposite in physical appearance, styles, and presentation. He had to make a decision about who would be promoted and who might need to be let go. He was gravitating toward the blonde.

What the hell is her name…Cratty, he thought as he lounged in his satin robe wondering what the day would bring him.

His only regret was his need to check in with his own boss, Chairman Eric Daniels, before he made his decision.

Webber had spent a great deal of time and effort learning as much as he could about his boss but to no avail. The man was shrouded in mystery. Daniels's career skyrocketed after the killing of Oman Sharif Sudani during his watch. After that, though, Daniels was a man in the shadows, typically conducting his business over phones or computer links. Rarely in person.

Still, Webber was not going to let his boss and job get him down today. Today was the third day of a two-week vacation. He had told people he would be abroad in Germany and incommunicado. Instead, he stayed home.

They'll never find me here, thank God. Now that I think of it, I think Eric's out for a couple of weeks too. Where does he go? Somewhere south I bet, Webber speculated as he went through his mental check list of pleasurable things to do.

Eating very expensive food and drinking lots of wine, his best treat was having a couple join him for dinner, a late swim, a dip in the hot tub, and then sex.

It doesn't get any better than this. I'm just glad they were able to keep their plans. And for a whole week too...now where did I leave that chocolate?

Having known the Barrys for years, it was their best kept secret. He would coordinate every vacation with them, even though it was sometimes difficult because of their own schedules, but it was definitely worth it. Fortunately, they had planned several weeks in advance.

Still exhausted from last night's romp, he strolled into his walk-in closet lined with floor-to-ceiling mirrors to see what would be comfortable.

Nothing. I think I'll stay in my robe.

Next, he decided that because it was such a beautiful day, he would go for a swim and a late lunch before he made such a weighty decision about the two candidates.

"You know, the dark haired one would have been a good choice but she's just too edgy. You'd think she'd learn how to relax. Find a hobby or get laid or something," he said as he walked to the kitchen in search of his missing chocolate.

"Wow! It's another gorgeous day out today! Just perfect weather!"

Looking in the refrigerator, Webber found the TV clicker and was about to turn it on when he finally saw his candy way in the back.

"There you are, my little friends! Now for some coffee..."

While still grasping onto the television remote, Webber threw it over his head where it landed on the couch where his papers, books and magazines were sprawled out.

"Nope. No TV, radios or computer screens of any sort. Just me communing with nature, coffee and, of course, my chocolate friends," he said as he carefully chose his first piece.

It had been beautiful weather and he was doing well by not watching the news or reading the paper today. He hadn't at all for the last two days.

"It's great to be me," Webber said to no one.

Catching himself in the mirror, he stroked his face, trying to decide if he should shave.

Nope ... too much effort, he concluded as he headed back to bed.

In Waltham, Massachusetts, hiding in plain sight, the operations center was buzzing with activity. With nearly forty monitors, three floor-to-ceiling screens, and sixty-three staff all in varying degrees of alertness, Jillian T. Davis, an operations center's manager, was moving from one bank of screens to the next. There were a total of four critical missions going on at the same time; all were abroad in hostile, foreign territories, but fortunately, they were mostly observations, reconnaissance, and intelligence gathering. There was one operation that might move from passive to active rapidly. Hence, Davis had cut her early workout short and came in long before her shift was supposed to start.

Looking around at the controlled chaos, Davis was wondering if it was a mistake coming in early.

Well, you don't need two of us here. Maybe I'll get a bite to eat and come back after the shift changes. Glenn can give me an update then, she thought as she slowed her pacing down to an amble.

Stifling a yawn, Davis tried to look less tired and present as she took her time scanning the ebb and flow of staff in the operations center. Fighting to stay awake these days had been more challenging of late.

Those damn dreams! What the hell are they about? You'd think by now after years of them I'd either be used to them or stop having them, she thought as she tightened her hands behind her back, away from her neck.

God, they're exhausting. It's like I never fell asleep. Wolves, birds, eagles, crows and shit...what's that all about? OK...think of something else.

It was unusual to have two managers on shift at the same time. The assigned manager was always in command, and the visiting or supporting manager would be second in command should the need arise. In regards to years and experience, Davis had the advantage over most of her peers. As a former US Navy lieutenant with years of field experience, Davis's presence naturally commanded respect. Six feet tall, athletically built, with very dark brown eyes that bordered on black, her physical features only added to the command.

Stopping in the middle of the center's core, Davis's hand found its way to her gold chain that held a crucifix.

Damn it, she thought as she yanked her own hand down from her throat.

If there was one thing she wished she could change about herself, she would get rid of that nervous habit. Whenever she was nervous or thinking, she would unconsciously fiddle with her necklace. She felt fortunate in that she had two sets from her mother; one was a plain gold necklace with a Catholic cross, and one just like it but silver which she kept at home.

In the past, she wore nothing that would constrain her neck or indicate attachment, weakness or human flaw. But these items were different in that they were her mother's crosses. While Davis had been closer to her father, he had died when she was an adolescent. Her mother, a devout Catholic, lived until Davis was in her late twenties. Davis's relationship with her was more complicated, as her mother did not approve of her vocation and profession. Still though, when she died, Davis had nothing but her mother's old home with very few personal belongings.

Except for this and that damn house, she thought as she pulled her hand down from her throat again.

This is worse than smoking!

Davis felt compelled to wear one of them every day. Pushing her thoughts to the present, Davis stopped her fiddling and looked around to see what was happening on the monitors.

While Davis did like being in command, she was not the manager of the first shift's Alpha Team; Denise Cratty was. Cratty was the antithesis of her: where she was quiet, calculating, and at times brooding, Cratty was gregarious, approachable, and engaging, engendering an esprit de corps. Some of Cratty's attractiveness was her physical presence such as holding a very feminine five-foot-five stature, light blue eyes and blonde hair that fell on her shoulders.

I bet your toenails are painted a bright pink. Now how did you make it through boot camp? For a manager, you're quite the "girly-girl." Did you even cut that hair when you actively served...oh, I forgot. You were in the Army. Still though, for a grunt, I'd expect a bit more command presence than Mother Earth, she thought while straining her eyes to see what was happening in Boston.

While Cratty was a former lieutenant in US Army Intelligence at the Pentagon, Davis held the respected position at the Office of Naval Research and Development. That gave Davis more training in life-and-death scenarios, while Cratty was more proficient in the political blood sports of Washington politics.

Well, there's very little guess work to see who's going to get that promotion. Now why did I even bother? Webber hates women who are taller than he. Especially if they can whip his ass in the ring...I probably shouldn't have done that, she thought as visions of being Deputy Director faded into the darkness.

While the rivalry between the military branches did not help their relationship, it was their vast difference in command styles that separated them the most: everyone wanted to work on Cratty's shift, and no one wanted to be on Davis's detail. Because the operations center was staffed mostly by civilians who were technologically ahead of their peers, there were no military staff in uniform present. Soldiers wanted a leader like Davis, not Cratty. For a mostly civilian operation, Cratty would pull on people's strengths and help them grow; Davis demanded the very best.

"And if you make it on Bravo Team, you are destined for clandestine field work and covert operations. After that, you'll have your choice working stateside for any private security firm," she would tell her staff.

That was very alluring as it was common knowledge that the money from those private firms was fantastic.

Maybe someday. But not today, she thought as she moved closer to the Boston/Merrimack Valley group of monitors.

Even though she stalked Alpha shift because one of the operations could go hot, she had been hearing a lot of reports about the Merrimack Valley. If she was running the shift, she'd have the shift on lockdown with no one coming in or out.

But this is Cratty's team...

Cratty had a history of always welcoming any manager – Davis, Glenn, anybody for advice. Davis had always wished she could feel the same way about Cratty; however, that was not going to happen, even though Davis had noticed that Cratty seemed more subdued and less enthusiastic in the past two weeks. The scuttlebutt was that Cratty's mother was very ill and that she was also having difficulty adopting a child. Davis had to admit that she felt bad for her. Davis started to fiddle with her cross again as she realized that she and Cratty might share something in common. Davis felt suddenly guilty for her earlier ranting.

I hope your mother gets better...

While four missions were happening in real time, there was a small team watching the events happening just next door. Davis was now hovering over this bank of monitors because she was truly disturbed by the events.

Bombs in a hospital, burning buildings, chemical agents in parking lots, and school evacuations...all tied to Arabic writing and Oman Sharif Sudani's death. I wonder which group of shitheads is going to claim responsibility for these attacks!

Chatter picked up communication that the FBI in Boston had gone dark as a result of a nasty cyber-attack. Davis knew Helms very well, and she greatly respected the Marine.

He's not going to let this attack go unanswered. He's a tough bastard.

The US National Guard, State Troopers, Coast Guard, Air Force base, all cadet and training academies, everything was mobilized to stop what appeared to be a domestic terrorist plot from unfolding. But that was not the thing that disturbed her most. Other than a shooting between federal agents, two members of organized crime, and a John Doe, there were absolutely no other injuries or casualties as a direct or even indirect result of these attacks.

This just ain't right. Complete news coverage and no injuries? Either this group is a bunch of pacifists or they really have lost their touch in inspiring terror. This runs completely contrary to the purpose of invoking fear through a large body count.

Looking at earlier transmissions of the terrorist's texts, she was also surprised with how it sounded more like an American-style, Navy Seals operation. Davis knew from experience that when there were strands of data that went one direction but ended up in the opposite place, there was always something bigger at hand.

"What do you think?" Cratty asked as she mysteriously appeared behind her. To her credit, Davis didn't jump even though she was engrossed in the data, and didn't see Cratty until she was upon her.

So much for my ninja-like reflexes, she thought.

"I'm not liking it," Davis responded as she turned to address Cratty.

Jesus, Denise! Hair's a mess, make-up is far from its usual perfection, circles under the eyes...You really do look like shit, she noted.

"What bothers you about it?" Cratty asked while looking around the room.

It was a genuine question. As much as Davis did not personally like Cratty, she admired Cratty's willingness to ask questions to solicit opinions.

"A sign of a good leader is having your people work for you and think it through, not you doing all the thinking for them," Helms would always say.

Her response was "I don't have time for babysitting."

Trying not to stare at Cratty's dishevelment, Davis focused on why she was suspicious of the whole situation.

"No injuries. No casualties. No body count. All the components of a full-scale terrorist assault with local, national, and global news coverage, and absolutely no one hurt. They even have an effective cyber-attack that took out the Bureau and could blind surveillance, but still, no bodies. How does that happen unless you carefully planned it that way? And who would do that? And if it's planned not to cause harm, who would want to do that on such a large scale, and why? Distraction?"

"There is another problem," Cratty added quietly as her eyes settled on Davis.

More than this?

Without waiting for a response, Davis looked blankly at her.

Hey Cratty, have you slept at all in the last 72 hours? I've seen cat shit that looks better than you!

"One of our senior field agents was scheduled to make contact with his team. He's disappeared, and his crew are MIA. He was also scheduled to come in for a debriefing at the beginning of the shift, and he was a no-show. I've sent a team to his home just to check up on him. The description of the situation in North Reading with that shoot-out sounds like there could be a connection. After we check the house, we're going to see the body and check the crime scene. Before the bureau went dark, they were lead on the shooting, so we're trying to get a hold of the deputy over there."

*Now this got real serious. One of our team is MIA? Who could do that? Maybe they just went dark themselves...*Davis hoped.

"Is that why you have this group monitoring the Merrimack Valley?" she asked.

Davis waited for Cratty's typical zen-like like retort but she was surprised when she noticed Cratty's attention drawn to one of the monitors that had brought up a transmission that had been immediately forwarded to the operations center.

For just a moment, there was one paragraph spelled out on one of the screens. Davis saw it and couldn't help but think it was a clear threat sent specifically to the operations center.

> *"I remember everything on May 2, 2011. Operations Center, Foreign Intelligence Agency, all present foreign operations are compromised. Cyber-attack imminent."*

On all the smaller monitors, there were telltale signs of a viral infection — a lock out, frozen screens, pixelation, and then a fade to black. Based on the FBI's cyber-attack, the difference was the speed of this attack, which seemed instantaneous.

Davis's attention went right to the large floor-to-ceiling monitor that was beginning to pixelate and break up. She was surprised at the speed of the attack as she watched the middle monitor go completely dark as the two flanking monitors were fading out. Immediately, there was a sea of exhales, sharp intakes of air, and a bunch of expletives from analysts and operators. Cratty was the first to respond.

"Shut down! Everyone, kill the juice to your computer now!"

Davis knew why she was responding so quickly. What had crippled the FBI's office in Boston was now targeting their operations. And because this was happening in the middle of four real-time operations, whoever was responsible would be held for treason.

Bastards, she thought to herself. *People could get killed!*

Some staff did shut their computers in time; more did not. For most it was a matter of watching frozen screens, and then they saw e-mails, files, and programs begin to vanish from their computers' hard drives.

Standing in the dark with only the sound of air-conditioning and breathing punctuating the silence, Davis watched Cratty look over her dead operations center as she clearly balled her fists in rage.

So much for a peaceful, sensei-warrior-like response. You look pretty pissed, Denise.

Davis waited for Cratty to invoke the emergency protocol.

What the hell else can you do? We're dead in the water and four missions are at risk.

"All right. Alpha Team, prepare to redeploy to Hanscom Air Force Base. We are heading to the bunker."

While not in uniform, they were in a fire zone and under attack. To Cratty's credit, she did exude confidence and command to her staff and to all who were visibly shaken, which included nearly all of them. Davis began to think that maybe she and Cratty were cut from the same cloth; their reaction to the cyber-attack was to strike back.

Soldiers in arms, she thought.

Finally Cratty turned to Davis, and officially issued the order to effectively abandon their posts and regroup at the fallback positions. There were always plans and protocols; people always drew comfort from that. Davis was struck by Cratty's relatively low tone but firm order.

"Davis, I need you to remove all critical operational files and data and to reestablish contact with all foreign and domestic operations. I need you to immediately dispatch to the auxiliary control room and keep things going until the bunker is up and running. There is a window of sixty minutes."

"Yes, sir!" was Davis's response. Davis knew Cratty's shift to military orders and protocols was done to demonstrate to the staff of civilians that they were all one team, and that as a team they would prevail.

After all, we are all at war with whoever did this.

As Cratty's second-in-command herded the staff out into the exterior offices and prepared them all for transport, Cratty and Davis walked to the manager's office, which surveyed the entire operations center. Once alone, Davis and Cratty spoke freely.

"This just pisses me off," Davis said as she readjusted her necklace so that it stayed under her blouse.

"Tell me about it. Who the hell did this?" Cratty asked.

"I don't know, but I'll get the auxiliary control going to cover the other missions," Davis responded.

"Good. That will give me time to find the assholes who did this," was Cratty's unusually blunt response.

Davis took a moment to consider Cratty as she was pulling documents together. Davis never thought she would ever hear Cratty swear.

Wow ... she's really pissed, she thought as a small smile formed.

Maybe I could get to like her. We won't be barroom buddies but she might not be too girly-girl.

As if to confirm Davis's thoughts, Cratty looked up and saw Davis looking at her.

"What?" she said with more than subtle annoyance revealed in her tone and expression.

"Nothing, Cratty. I just agree with you. Find them and kill."

"Yeah—" she said as she shifted focus to collecting documents.

"Assholes. You think you can screw up my ops center and just walk away? I don't think so," Cratty muttered as she unlocked the wall safe to extract four two-inch-thick operations and code books. In addition to this highly classified data, there were four external hard drives, each with five plus terabytes of data documenting five years of covert missions and operations. Cratty was listing off the inventory so Davis would have everything she needed to cover the operations.

While Davis was listening, she was mostly focusing on unlocking her assigned bin, taking out her military-issue, .45-caliber handgun with additional magazines and a retractable baton. As Cratty loaded the black backpack with the critical information, she also produced the key for the magnetic lock to the auxiliary control room vault and handed it to her.

Davis was intimately aware of the rationale of having an off-site base of operations as they reestablished the secondary command center. The auxiliary control room was an abridged operations center that could also be kept hidden in plain view while not connected electronically to the main center. This arrangement made a cyber-attack on the operations center moot.

I guess this set-up was a good call. Never thought it would come down to this, she thought as she readjusted her weapons, magnetic key and backpack for travel.

In addition to being separated and safe from attacks, it was small enough to be protected by a small detail. It could also be operated by one person. Four managers, two assistant managers, and the director knew about the location, its purpose, and the contents of this room, which kept things safe and classified.

Speaking of the director, where the hell is Webber? Someone has to know how to reach him!

Davis's mission was simple: take the car via the prescribed route, provide the security detail with the day's code to gain entrance to the room, and set up operations within sixty minutes. Because all members of this elite group had practiced this run and alternative routes once a week, and at varying times of the day, Davis could honestly say that she could get the base up and running in forty minutes.

Wearing a long coat to better conceal her weapons and stay blended into a civilian background, Davis watched Cratty as she rechecked her clipboard and checklist to make sure she had everything.

For maybe the first time in three years of working together, Cratty extended her hand to shake Davis's. Davis took it and was pleasantly surprised that it was much stronger than she had expected. "Good luck," Cratty said.

"Will do," Davis responded.

Davis took inventory of Cratty as she prepared to leave: with the exception of Cratty's mother's illness, Davis felt there was little in common between them. But today was very different - Cratty acted the way she herself would have if she were in command and there were a breach of this magnitude. Whoever breached the operations center was a danger to national security. Davis found another thing she had in common with Cratty — *she takes national security pretty personally.*

Davis was walking out the door when she turned back to say something to her just as Cratty was retrieving her own semiautomatic weapon and slapping a clip of ammunition in place.

"Cratty?"

"What?" Cratty asked with less annoyance than before.

"Find these punks, will ya?" Davis called back.

"Didn't we just go over this?" she said as she went back to looking for her other ammunition clip.

Davis couldn't help but smile broadly as she headed to the garage.

I really hope she finds those bastards. God help them if she does, she thought to herself.

Chapter 16

"Absit invidia"
"No offense intended"

Present Day – May 2

BURNS HAD PARKED THE ambulance toward the back of the private bank in what Samantha referred to as "launch mode," which meant it was facing out, ready for someone to put it in drive and hit the road.

As casually as he could, Burns started walking toward the back entrance. His scalp and hands were itching again, but he was sure it was just nerves this time. Based on the floor plans and David's surveillance tapes, Burns knew exactly how many stairs he had to climb to the second floor, where the security camera filmed, the front desk was situated, and the first guard would be standing. He also knew the entrance to the auxiliary control room and the most likely place the other guard would be stationed.

Burns was now using a paramedic backpack to keep both of his hands free. When he was halfway up the flight of stairs, he held his Taser in his left hand and a stun gun in his right.

Crossing the distance to the guard sitting down would be easy, but simultaneously shooting a Taser ten to fifteen feet at a target with accuracy while being up close and personal with another guard getting up would be tricky.

OK. Almost five years in the making and it's time, he thought as he released a sigh and rolled out the tension in his shoulders.

As it was a private bank in a public setting, it was not unusual for the guards to interface with the public, even though they were upstairs and away from most people. Today would be a little different. From their perspective, they would probably be wondering why there was an additional guard or law enforcement support there when they had not requested one. Burns was sure that they'd wonder why this man had a paramedic bag on his back. And that was exactly what happened.

That should give me my ten seconds to take them down.

As Burns rounded the corner with the exact steps measured out, he saw the first guard behind the desk clearly reaching over to either pick something up or tie his shoe. The second guard directly behind him had his arms folded across his chest, presenting Burns with a more compact target. The standing guard was struck first with the Taser shot, which flew right over the first guard's head. It hit its target, shocking the large man and sending him into convulsions as the voltage ran through his body. The seated guard witnessed this and then turned around, his semiautomatic weapon drawn from his holster. But it was too late. He was now convulsing from the shock from a stun gun.

In less than ten seconds, both guards were down. Because the operations center would be off-line, the feed from the security cameras would have to be viewed manually from the bank. This meant that a live response and a strike team would not be sent. The only person he expected to show up next would be the designated operator of the control room.

OK. Now we wait.

He had maybe ten to fifteen minutes to stage the area so that it looked like he was assisting a downed guard and hopefully get the jump on the assigned field agent.

Just as Burns was getting the paramedic bag unpacked so that the scene would have that "crisis look," he heard Becky's voice over the radio:

"Tiny to Falcon. Scarlet is in place, and I have eyes on target car. It's heading to the right of the exit. Repeat right. Plan B."

So much for precision. I really didn't want a smash and grab, he thought as he kept working.

It had taken years to find the operations center. It had taken weeks to find the cars transporting the managers to the auxiliary control room. Finally, the end was coming.

"ETA?" Burns asked.

"At that speed, five minutes."

"Okay, Tiny. Do you have eyes on Scarlet?"

It was a long minute, and Burns worked through it, securing and hiding one guard beside a small leather couch.

"Yes. Eyes are on Scarlet," Becky replied.

"As soon as she is halfway there, blow the generator and set everything at the PD in motion. Pack the gear and go to rally point delta. Falcon out."

Burns was beginning to sweat now as he looked around and saw that the guards had the same shirts, insignias, and badges. That gave Burns another idea to improve on the original plan to pretend to be an assisting first responder. He hoped it would work. Burns moved ahead with his new idea, but he wished he could make sure that Samantha was all right. He knew she would be fine, but he was just anxious about her well-being. Burns unexpectedly smiled to himself as he tried on his new shirt.

I just want this over so I can be with her soon, he thought.

As Samantha parked her car, she bent over to ostensibly tie her shoe. In reality, she tripped the receiver for the high-yield explosives near the gas tank toward the back of the car. She made sure not to park too close to other cars in the "police only" parking lot. This was easy to do as the station was sparse with everyone out on the streets.

The civilian parking lot, on the other hand, was jammed up beyond capacity. Everyone wanted to get information about what was going on in their community as sounds of sirens, fighter jets, and helicopters could be heard everywhere. As she walked with purpose toward the front doors, the doors of the police station suddenly opened, and they were held open by officers to let the civilians out of the building.

Attitude, Sam...it's all about presentation and attitude.

But then, she remembered she was supposed to be a cadet.

No wait. It's "compliance" time, she thought as she realized that she would be on the bottom of the command structure.

One of the officers looked over and called her over.

"Cadet! Either hold this door open or get in there and help evacuate the building."

"Yes, sir," was her response as she ran into the building. While in motion, she put her radio earpiece in so that she could talk to Becky.

Samantha asked one word, "Generator?"

Becky responded, "Yes. Ready for the rest and then mobilizing to rally point delta. Falcon is onsite."

That's it. This shit will be over in about an hour...

As the crowds were moving by her and as she guided them to the front door, she made her way to the completely unguarded front desk. While she went against the tide of civilians, Samantha went completely unnoticed as she navigated to the stairwell leading to the interrogation rooms. As she descended one flight, she tripped the switch to ready her transmitter. Before she walked down the flight of stairs, she stood near a ground-level window and spotted her car. Well within the line of sight and still no one around the car, she tripped the button from "ready" to "fire."

She was about to hit the button when she looked out one more time to make sure the car was clear of people.

"What the hell...where did she come from..." Samantha said quietly to herself.

Samantha's heart stopped as she saw a white ball bounce toward her parked car and get stuck under the front wheel. She immediately looked down to make sure her thumb was nowhere near the trigger. As beads of sweat burst on her forehead, she watched a little girl run toward the car to retrieve her ball. She was a little older than Emma.

Jesus, she thought.

Samantha felt as if her bladder was going to burst as the little girl wedged herself under the car to get the jammed ball. Samantha watched, frozen in place for fear that any move she made would trigger the explosives. She watched helplessly as the little girl squished herself under the car and pulled at the ball. Suddenly, a woman, probably her mother, and a police officer came to the little girl's assistance.

"Shit, shit, shit, shit—" Samantha muttered under her breath.

As the woman pulled the little girl up from under the car and started scolding her, the police officer bent over to pull the ball from under the tire. Instead of bending to look under the car to see the ball, the officer squatted down to feel for the ball.

Samantha was barely breathing. Then she took a sharp intake of air and held it. Unable to feel for the ball, the officer dropped to the ground and looked toward the front of the car, the woman and little girl waiting patiently. Samantha closed her eyes with a vision of the bomb being about five feet from the back of his head.

If he turns around, he'll see the explosives. What will I do? Trip it anyway? If I don't, he'll warn everyone, and we'll all get caught. Damm it!

With a life-and-death decision in play — press the button and kill them all so she, David, Becky, Emma, and Burns could escape or spare these strangers and risk losing her loved ones - Samantha found herself confused and scared.

Samantha opened her eyes to see if her fate or theirs had been sealed.

I don't know, she desperately thought.

The officer never looked behind his head as he immediately found the ball and pulled it out without an issue. As he stood up, he threw the ball to the happy girl, who held her outstretched hands to grab it. The woman and little girl kissed the officer.

"Oh, shit. They're a family," Samantha muttered to herself.

As the family started to amble away from her car, she anxiously watched them retreat. As the distance grew between them and the car, Samantha started to breathe again while she continued to watch to make sure no one else got close to her car. Once Samantha saw that the family was out of her line of vision and no one else was around, she pressed the button before anyone else tested her moral character. One second later, there was a massive explosion in the parking lot, one made more spectacular by the explosives' proximity to the gas tanks as well as the additional gas in the trunk.

Good Christ! What is he using for explosives!

Samantha felt pools of sweat under her armpits, and she immediately used her sleeves to wipe the perspiration off of her face. Leaning heavily against the wall, she produced her radio as she watched the blaze of fire consume the car. Her throat was dry, but she was able to croak out her message to Becky.

"Scarlet to Tiny. Backup," which meant that she needed to get rid of the emergency backup generator.

"Done. Love, Tiny."

Love you, Becky, she thought to herself.

Just then, the remaining emergency lights flickered out, leaving everything pitch black with only the light from the ground window providing any illumination. As Samantha waited a moment to let her eyes adjust to the darkness, a civilian was stumbling down below in the darkness. Samantha jumped back into character.

"Are you okay?" she asked.

"Yeah ... I got to get upstairs. Can you get the lieutenant and his prisoner? He is in one of the interrogation rooms," the civilian, maybe some kind of computer guy, said as he focused on getting out of the building.

"No problem," Samantha said as she helped him up the stairs and out to the lobby. Once alone, she took out two flashlights and headed to the target room. Her final objective of the mission was to liberate David. She was sure her acting as a cadet would be believable. Playing a part or a role was second nature to her.

With Becky's part as the lookout and trigger man now over, she began the process of wrapping things up. She had all the laptops packed, all gear and clothes stowed away, and all she had to do now was get the car loaded, and get Emma and herself to her rally point. That was where Burns, Samantha, and David would rendezvous, split up any cash, data, and files, change cars, and separate. David, Emma, and Becky would drive to Bethlehem, New Hampshire, where their off-site storage was located, exchange the car with Massachusetts plates for a family minivan with Canadian plates, and cross into Canada. With Canadian passports, they would act as Canadian citizens who were "escaping the American terrorist attacks," and returning home to Montreal. It would be two and a half hours to Bethlehem and then three and a half hours to Montreal from there.

Once the car was packed and Emma was in her car seat,

Becky debated turning the radio on to hear the news. She decided that knowledge would be better than any surprises at this point.

The station she chose was just "concluding the emergency broadcasting alert" before it cut to a news report in progress. Still, the newscasters were alerting the public about efforts to evacuate key areas of the Merrimack Valley. The newscaster spoke of explosives, fires, cyber-attacks, and an assault on federal offices. The attack was being highlighted with reports about computer virus protection companies and how they were locking down and shutting off all their products as a result of a nasty worm called *Albatross*.

"Not only does this worm piggyback on the virus protection definitions, it then freezes and erases your hard drive of files and e-mails. Additionally, it changes your password, leaving you without any way of accessing your computer again," one newscaster was wrapping up.

Another chimed in with how the worm was affecting anything that had any connection to and use of commercial virus protection, including the "cloud software" for remote access to computers which was once thought to be invincible. The second newscaster summed up the degree of damage in a few sentences:

"All in all, this virus could easily affect, disable, and disrupt all personal computers in more than 70 percent of all households in America. That could be as many as 217 million computers. Military computers have their own separate system of protection, leaving them operational, but the president has put all domestic and foreign bases on the highest alert. Many critical systems at hospitals and airports seem to also be unaffected, though there is question if any fatalities may be a direct result of this attack. Early reports suggest that this attack may be a coordinated effort between Islamist fundamentalists, a North Korean splinter group, and possibly a domestic terrorist cell."

Jesus ... did I kill someone? That's enough, Becky thought as she turned off the car radio.

"God, I hope no one's hurt. If we're caught, we're really fucked."

Becky panicked at her outburst, and she looked back to see if Emma had heard her swear. Fortunately, Emma seemed to be preoccupied with her handheld game. Becky sighed.

The last thing I need is for David to ask me to please watch my language as the whole world is thrown into chaos. He would, too, I know it.

Still, the thought of someone, maybe a child, being hurt was sickening to her. Becky did think that if she did cause someone harm, going to jail might be better than living free with the guilt. Her mind suddenly jumped to how her sister was able to live with her actions. Becky knew that Samantha had killed at least two people in self-defense, but she was not sure if she could live with it as well as her sister could.

Could I really use the same excuse? Self-defense? Becky thought. She knew herself well enough to know that the guilt would just kill her.

Becky tried to push the thoughts out of her head and wondered if it was too soon for another anti-anxiety pill. It was pretty surreal to think of herself as a terrorist in the morning and then to become a Canadian citizen having dinner, sipping wine on St. Catherine's Street at 7:00 p.m. that night.

"I can't wait for this shit to be over," Becky said quietly.

"'Shit!' Mommy said 'shit,'" Emma belted out, followed by loud laughter.

Closing her eyes, Becky pictured David's disapproving face.

Shit, she thought as she tried to ignore Emma's enthusiasm about catching her mother saying a bad word.

Jail's better than that look he gives me.

Chapter 17

"Falsus in uno, falsus in omnibus"
"False in one thing, false in all"

Present Day – May 2

ANDERSEN WAS JUST OUTSIDE the interview room, and was getting an update on the investigation when he reiterated his belief that David had been part of the crime and was not just a witness.

No. This is not a case of a person being in the wrong place at the wrong time.

It was now close to 11:00 a.m., and in addition to the shootout with federal officers, a member of organized crime, and someone who was still unknown, there was a "smoking" truck in the town next door, and a school in Andover that was under a bomb threat. To top things off, his boss was no longer deploying resources to his town and the neighboring town of Lawrence, but was now in the middle of coordinating all resources with federal law enforcement agencies for the Merrimack Valley.

Now how the hell are we going to do that? We're a small town, not an urban PD.

The last piece of the "update" was that a convoy of FBI agents was en route to Andover to assist in the evacuation of a five-block radius while the governor had now deployed the US National Guard, Coast Guard, and all police departments and academies to assist in what the news was calling "the Merrimack Valley Crisis."

What really stunned Andersen was word that the regional FBI field operations were completely shut down by a cyber-attack.

Well, this just became a federal case. I'm guessing this is now an act of war...Are you responsible for this, Caulfield? You and your friends, Andersen thought as he sketched out his next move.

In conclusion to Dempsey's report, he informed Andersen that there was nothing but a skeleton crew upstairs, and that all the prisoners were fed earlier than scheduled and were now in lockdown until more resources became available. The second and third shifts were also coming in early, and the Red Cross was out with the other first responders setting up food stations.

"Oh, yeah, the counselor's assistant showed up. She told the boss she was not able to access the files he asked for and needed to go to the office and bring back hard copies," Dempsey said.

Andersen couldn't stand it when Dempsey would leave the most relevant piece of data at the end.

Why does he always do that – leave something out? Why can't he just give me everything at once?

"Did you get her name or what she looked like?" Andersen asked as he pushed his negative thoughts about the officer out of his head.

"No. The boss spoke to her. She was pretty frazzled, and she had her kid. It got pretty busy upstairs. She was only there for ten or fifteen minutes or so."

Andersen thanked Dempsey absently as he closed the interview door. Deep in thought, he walked around to his empty chair trying to find the right words to start the next phase of the interrogation when David answered his nagging question.

"Don't worry. It wasn't Becky and Emma."

Andersen was beginning to dislike Caulfield with his crazy ability to predict thoughts and questions. Staring at him, Andersen felt the corners of his eyes narrow, his chest tighten and his fists clench.

But again, that was what a therapist should be able to do. But I'm tired of this bullshit.

"Enough!" Andersen said sharply and hit the table with his now open hand.

Amazingly, David did not flinch.

What?! Did you predict that I'd be frustrated enough to bang the table, too? Did you expect that and that's why you didn't flinch? Sitting quietly for a moment, Andersen concluded that David was not only involved but that he was dangerous.

"How do I know you're telling the truth? How do I know you're just a witness and not part of the problem? How do I know you're not orchestrating everything right now?" Andersen questioned.

"Because—" David answered casually. "Becky and Emma left me four months ago."

Andersen stopped cold. After an uncomfortable moment elapsed, David went on without prompting to fill in the gaps.

"Once I was set up with assisted living, Becky and Emma left. I think Samantha is dead, and I hope Burns is either dead in that house or you find him and kill him. If he's not dead, then he is more dangerous than ever."

Andersen was quiet. David was now telling him that this team of terrorists was no longer in operation and that we were now dealing with just one rogue operations specialist. Andersen was still trying to put it all together when David leaned forward again and began to explain.

"This is what was supposed to happen. Here was the plan. We got four adults. One is a specialist in the area of destruction and mayhem, and three civilians have nothing to lose and all to gain. Once we found the operations center, we had to watch everything and everybody coming in and out of the building. We were looking for patterns, people, and routines that might lead us to the objective."

"I thought the objective was to find the operations center," Andersen interrupted.

"No. The objective was to disable the operations center and force them to relocate to a place called the auxiliary control center as they shifted all programs, operations, and missions to the bunker or something for complete security lockdown," David answered.

Even though David was blind, he could tell by the silence that Andersen was confused, so he continued.

"Burns said that he believed that if the operations center was compromised, there would be a seasoned field agent or manager immediately dispatched to a secure area in the community no more than ten minutes away to carry on critical operations, communications, and investigations for the three hours needed to get the bunker to full speed. The backup auxiliary control center had to be up and running in sixty minutes or less."

David stopped, took a breath, shrugged his shoulders to release tension and then went on.

"Of course, while finding this operations center took years, I thought finding the people or person and location of this smaller control center was going to be harder. But then I underestimated the skills and resolve of both Burns and Samantha. They were constantly out and working incredible hours. Becky got involved for more than a month. And then I guess it was Samantha who noticed a pattern of two people who fit the criteria Burns laid out for what to look for, including the entourage security around some guy. There was a woman who was with him who was clearly not a date, girlfriend or peer but a subordinate. She was a very seasoned professional Samantha described as 'scary,' and Samantha did not scare easily. Once the field of possible people was narrowed, we got working on following and locating where they went during the day. That woman we thought might be the person — we called her Cougar — was the right call."

Andersen had to stop him.

"You called her Cougar?" he asked.

David's hands immediately flew up in frustration as if he had been repeatedly questioned about the name, its double meaning, and was simply sick of it. This was the only time he saw David flustered.

"Look! I don't know what she looks like, but they told me she looked like a cat ready to pounce. I said the first image I had, which was 'she paces like a caged cougar or a lioness about to leap.' Apparently, I missed the double meaning of the 'cougar' statement, and so she was forever referred to as Cougar."

David still sounded embarrassed and apologetic. As if not to dwell on the embarrassment, he went on.

"Once we had the possible locations of the operations center and this control room, we had to come up with a plan to compromise this place and to create as much collateral damage so as to confuse the situation and provide cover. While Burns was positive the operations center thought he was no threat and alone, he also thought he needed to have a simple plan to create a crisis."

Andersen saw it clearly now.

But he would need help. He can't be in four different places nearly all at the same time.

"So Burns set the building to explode, created a crisis at the hospital and the parking lot, and then the school? But how could he be in all of the places at the same time?" Andersen asked.

Unless Samantha and Becky are actively assisting him. If so, why would they leave you? What's gained by leaving you in the middle of a shoot-out? Did something go wrong?

"Burns had us set up everything in dry runs, so staging it and timing it was easy to do. All you need is five or six able-bodied people. He had two women, a blind guy, and a baby. Not exactly an elite strike force. But the problem of getting hired guns, I guess, would be pretty easy. Money talks, and we had that for small jobs like 'set this up' and 'park here' and 'watch this.' One thing that worried me, Becky, and Samantha was why Burns did not want us to know the location of his ultimate target. Becky had to convince Samantha that finding out this location was important for our safety."

David stopped. It was apparent he was pained by something. Andersen had to be careful here.

Shit. What happened now?

"So what happened?" Andersen asked gently.

It took David a few seconds, and then he went on.

"Samantha, Becky, and I had our own code words. Code words just between us. Because Becky and Samantha were sisters, they had their own shorthand and communication between them. I didn't mind because I needed them and I was not a threat. I always got the impression that Burns let Samantha into his life as a way of gaining access and control. Because Samantha was hungry for this affection, she really wanted to believe. I think Becky never trusted Burns and reminded Samantha that her first loyalty was to her and Emma and that I was second. Burns was a distant third."

David slowed down again, but he kept going. It was hard to say, but Andersen thought David's eyes had welled up.

"Samantha and Becky had to get their hands on a federal laptop to make an easy connection to the operations center. The FBI was the easiest federal agency to find. Finding out where the agents would eat, drink coffee, or just drink was also very easy. Samantha was the one who lifted the laptop while it was still open and on while one of the agents was in the bathroom and the other was assisting Becky and Emma out the door with a carriage, bags, and coffee. So getting the laptop was really easy. But when Samantha went to give Burns the laptop, she texted Becky an important code word, 'marathon.' If any of us used the word, that meant anyone hearing it was to take everything they could and run. 'Marathon' meant Burns was done with all of us and that he planned to get rid of us."

David took another breath and slowly went on.

"Becky, Emma, and I already had a go bag packed and cash, so it was easy to just leave. Becky had a method of e-mails and texts to contact Samantha, but there was never a word from her again."

Well...I guess this Burns guy is a cold, heartless killer.

"Burns killed her?" Andersen asked while silence blanketed the room.

His witness looked off into the distance, shrugged and shook his head. It looked like he was trying to find the right words before he answered.

"I would like to think she got away, but if she did, she would have contacted Becky. She never did. After one month of hiding, it made sense for Becky and Emma to get far away from me as I was simply slowing them down. It took a while, but I convinced them to leave me."

David was now dark, Andersen noticed. The words "sad" or "melancholic" did not do justice to how he looked.

"I lost another family," David continued.

"But I couldn't let anything happen to Emma and Becky. I did not want to be responsible for their deaths too. I was able to stay low for a while, because finding the limited number of assisted living housing facilities in Massachusetts is difficult. But not for him. Even though I was now working and needed an assistant, I probably made it easier in the end to be found. I'm guessing he left me alone for a while until he needed me for something like a fall guy or something.

Burns might have been waiting for Becky and Emma to visit or come back, and then take us all out."

"So how did you end up in North Reading at a shoot-out?" Andersen pushed gently.

Finally, the whole thing is making sense, Andersen thought. He had to have had a full notebook of comments so far.

This is going to be one hell of a long report. No one will ever believe it. It's just crazy, the whole thing...

"I was heading home last night when I was grabbed. Three men. One was Burns, and the other two were some pretty strong guys. They had brought me somewhere, and I ended up going down a flight of stairs into a basement, I think. I sat there for a while and tried to figure a way to get out. At some point, there was a bunch of footsteps coming my way. I figured I might as well go out fighting, but I was whacked from behind, and the lights simply went out. I was kind of in and out, but there was clearly some shouting, and I swear I heard someone yelling, 'FBI,' and then lots of shooting. But then I must have passed out again. Next thing I knew, I was found by police and just thrilled to be alive."

David fell quiet again.

Three guys? There were three guys at the house but if one was supposed to be Burns, wouldn't there be four guys at least? Hm... something doesn't add up.

"So what was the laptop for?" Andersen asked as a way to buy time so he could try to figure out what was wrong with David's statement.

"I heard Burns talk about the need to test some kind of virus or worm to simulate an attack. Something small just to see if it would work. Once he was convinced that it worked, he would wait two weeks to set up a bigger one with a lot of other events to confuse things."

What? Why would he wait?

"What do you mean other events to confuse things?" Andersen asked.

David looked puzzled and thought about it.

"I never really knew what he meant by that. But if I know the guy's thinking, he would not hesitate to use explosives, chemicals, or other weapons of mass destruction," David speculated.

Wait a minute. Who is your assistant? The one with the child?

"What did you want your assistant to bring you?" Andersen asked suddenly.

Obviously confused, David answered.

"I wanted her to access my e-mail about files and data on Burns. Why?"

"I don't know," Andersen said and then suddenly shot out the question.

"Are you aware that in addition to the shoot-out in my town, there was a bomb threat in a hospital and a parking lot of an athletic center in a residential community, and a cyber-attack launched at a school?"

Andersen thought David went pale and seemed stricken with fear.

David was about to say something when it suddenly seemed to Andersen that the fluorescent lights overhead brightened, surged, and then went out. The emergency power came up a moment later, but the smell of burnt circuitry and melting wires was obvious. David was the first to respond. It was evident he had heard something, and he definitely smelled the wires.

"What happened?"

"I'm not sure," Andersen answered.

Instead of a knock, the door opened quietly behind David. Andersen reached for his holster and realized that his gun was locked in his desk since guns were not allowed in the interview room. Andersen was greatly relieved when he saw it was the recording technician on the other side of the glass. Something wasn't right with the technician though. He really seemed confused and distraught.

"Sorry, Lieutenant, but there was a massive surge, and all the electronics are fried," the technician started.

"What does that mean?" Andersen asked.

"Well, it means that anything with a transistor, solid state, CPU and computer components are either burnt out or fried," the technician explained.

"I need to get upstairs to see if the computer terminals are all right."

"If the computers were on, does that mean they are damaged, too?" Andersen asked.

The technician gave Andersen a "what do you think?" look but still answered the question civilly.

"Sure does. Weird thing, though, was that everything started to freak out before the lights burned out. Like for thirty seconds or more. I would have thought that it was just a power surge like a transformer blowing out, but all the stuff would have happened at the same time. It didn't, though."

Andersen had more questions, but the technician was gone. There was a whole lot of noise outside because the door was ajar now, but the sounds were more muffled than clear.

"I hate to say this," David started, "but I've really got a bad feeling about this."

Undeterred, Andersen sat back down and started with another bout of questions.

"Do you think Burns may have already implemented his plan?" Andersen continued.

David seemed thoughtful for a moment and then seemed to draw some conclusions.

"Burns is many things, but his strong suits are details, timing, and logistics in general. If all of these events are occurring at the same time, near the same location and there was a cyber-attack, I am guessing this is a practice run. I am guessing he is timing everything, watching response times, seeing how things fall out. I would guess that he would more likely minimize casualties so that all of this would not draw a huge manhunt."

Andersen had another epiphany, but first, he needed to know the answer to one more question.

"What do you mean 'minimize casualties?'"

David looked puzzled at first but then answered the question as well as he could.

"It's one thing to have all of these events occur and have little to no lives lost but quite another to start taking lives and leaving a large body count. If Burns is successful in this practice run, he has embarrassed law enforcement at the very worst and tied up resources while determining strengths. He really doesn't want to drive the operations center back underground. If there are bodies strewn all about, then the United States of America goes to war, and everyone wants justice. Now we are talking 9/11 again."

"What?"

Something is wrong with his logic. If you do all this, you've tipped your hand. There's no surprise in the attack.

Andersen knew enough to know something was wrong, but he needed time to figure it out. At the surface, the reasoning made sense, but somehow, this did not fit the person David had described as a mastermind.

Clearly hearing his silence, David attempted to explain.

"When the Twin Towers fell, the United States went to war on multiple fronts with the backing of the US citizens and much of the world's willingness to go along. If Burns leaves a shitload of bodies around, he will have everyone in the world after him, and no one will rest until he is dead. Just like Oman Sharif Sudani and his buddy Bin Laden were hunted down and eventually killed in 2011. Burns doesn't want the world after him. He just wants to see how all of this works before he is able to get back at his old bosses for what they did to him."

Andersen had to think for a moment.

Was this all true? Could one guy do all of this? If Burns is that smart, a "test run" would show everything for nothing, really. He doesn't seem to be the type of guy to make an amateur mistake like this. Still, even Sudani had a team of people who field-tested strategies before they actually tried them, Andersen thought.

As Andersen thought more, he remembered that some of Sudani's most effective people were civilians converted to his cause.

It was when he was thinking of Sudani that another thought jumped forward.

What a minute— That name!

Andersen knew the name as did every American.

Hadn't Sudani been killed on May 2? How the hell could I have missed that? he thought.

Andersen was about to ask the next question when all of a sudden, there was a muffled explosion. It must have been a powerful one because it had made it through the soundproofing in the basement. Then the emergency lights went out, leaving him and his witness in total darkness.

"Fuck—" was the only word Andersen could muster now that he was in the darkness.

"Other than the explosion, did something else happen?" David asked.

"The emergency lights are gone now," Andersen said.

He also heard steps and saw two beams of lights outside the door and someone calling out for him. It was a female voice.

"Lieutenant Andersen? Lieutenant?" the woman called out.

Andersen was now up from his chair, and he placed David behind him so that whoever was at the door would not see his prisoner first.

"Sir! Cadet Smith reporting to you, sir!"

The body of the voice seemed to stand at attention but continued.

"The watch commander told me to report here to watch the prisoner. He wants you upstairs and armed. The station is under attack, and the prisoners' doors are no longer secured, sir!"

Andersen visibly relaxed for a moment. The young cadet seemed ready "to serve and protect."

These cadets are always ready to jump into the shit.
"At ease, cadet. Is that an extra flashlight?"
"Sir! Yes, sir!"

Andersen was amused and relieved. As he took the flashlight, the cadet took out her handcuffs to put on the witness. Andersen stopped her but directed a question at David.

"Are you planning on running, David?"

"Where would I go? I am safer here than anywhere ... or at least I thought I was. Out there, if Burns is behind all this, I am on the short list to remain alive."

The cadet objected.

"Sir, protocol requires that prisoners need to be secured at—"

Andersen cut her off.

"He's not a prisoner but a material witness. Just sit with him." Andersen turned to retrieve his notes, which were spread on the table in piles. He had a system of cross-referencing statements. He would need time to collect them.

The cadet seemed to pick up on Andersen's dilemma.

"Sir! I can secure your notes if you like sure, sir."

My God, the training about respecting rank is surely working.

"Thank you, cadet. Just sit with the witness and don't touch them. They are in a particular order. Just make sure they stay where they are," he instructed.

Andersen turned to David and tried to reassure him.

"Don't worry. We'll get him."

David said nothing.

Andersen left his only witness in the dark with a by-the-book cadet holding a flashlight because his police department was now under attack.

"What a shitty day this turned out to be," he said to himself as he bolted up the stairs to help with locking down the station.

There is no way Laura will ever believe this tale. Who would?

Silent, both David and the cadet remained standing in the dark interrogation room for a few moments until Andersen's footsteps were no longer audible. When David was sure it was just him and Samantha, he had to ask her a question.

"Have you done that cadet thing, role play or something before? You were too convincing to just pull that off without practice or something. You were just too good at it to be a first time."

David was relieved to finally be with someone he knew. It had been difficult to be interrogated by Andersen.

The man has excellent interview skills. He would have made an excellent forensic psychologist, David thought.

Eight months ago, Burns gave him and Becky computer sites, government websites, local papers, and open sources to research past and present law enforcement and military personnel that were in a twelve-mile radius of their target location. With Burns's specific search parameters, two names consistently came up as "very active" in law enforcement with history of military in the war on terror: Steve Andersen and Diane Welch. Much to David's surprise, past newspaper articles produced the most data. Still, Burns made it clear that there could easily be other people who were not in the public eye that could be a problem. David did appreciate Burns's caution. After some digging, Burns and Samantha found out that Welch was out on leave for "personal reasons." That left Steve Andersen as the only known variable that could cluster their plans.

While David could not "see" Samantha, it was easy for him to imagine her smiling as she took his arm and guided him to the door.

"You have no idea how many things I have been to so many for so often," she replied.

David pondered her response as it sounded very familiar...*historical even...*

Ah, yes ... the Battle of Britain, David recalled.

"Of course. 'Never was so much owed by so many to so few.'"

David heard Samantha's sigh.

"Why Becky finds your loose associations cute, I'll never know. But then she was always the smart one," she concluded.

But then Samantha added, "That was from Churchill's 'Finest Hour' speech, right?"

David knew she wouldn't let it go.

"You see ... you mock me, and yet you're quite the historian. You are right though. I should know enough not to ask you leading questions."

"You're right. You are pretty smart, and yet you do ask the most obvious questions. Have I ever played the 'submissive cop' role before?" Samantha said with a chortle.

David felt bad even casually referencing Samantha's former profession. He was uncomfortable with it.

I must be anxious, he thought.

As they moved away from the room, David stopped and pointed to the desk.

"Before we leave, we need to take his notes. The less he has, the better," he said firmly.

David felt bad lying to the lieutenant. While he did tell him much of the truth, it was all for misdirection and deception.

I'm just not cut out to be a con man, he thought.

David heard Samantha pull the papers all together into one big pile and stuff them unceremoniously into one folder under her free arm. They both moved out of the room and stood in the darkened hall. Being used to the dark, David was the first to speak as he pulled up blueprints in his head of plans he had long since memorized.

"So where are we?" David asked.

"Interrogation room eight," was the hushed response.

"Turn left. Twenty-five feet, there is a door that leads to a flight of stairs. Go downstairs, and it leads to a locked emergency door. Without electricity, we should be able to open the door in silence and walk out the back of the building. Cameras and detectors should also be off-line."

Well...I hope they're off-line. What if they have backups to backups? What if we missed something and the cameras or detectors have a power source we missed or located somewhere else?

"A walk in the park … how nice," Samantha whispered.

Let's hope, he thought.

With the directions perfect, they were now at the emergency door where the moment of truth was before them.

Will the other side be empty or filled with armed police waiting? Did we really pull this off? Did Burns get what he wanted?

Before they opened the door, David took his suit jacket and tie off and unbuttoned his collared shirt.

Not much of a change, but different enough at a distance not to be spotted as the "blind guy in the suit," he thought.

David heard Samantha make a more substantial change as she shed her cadet jacket and hat along with a whole bunch of stuff he couldn't identify. He guessed one of the items was a wig as the strands of hair hit his face.

"Sorry," she said with a short chuckle.

Finally done, they both waited a moment. Then David heard Samantha open the emergency door. After a remarkably loud and long metal creak, he was relieved to hear noise of people milling around outside, but no alarms blaring or armed men telling them to halt. With the crowd moving in generally one direction, it was easy for David and Samantha to walk arm in arm at the pace of the crowd. After a moment, he reminded himself that he could now breathe.

David was happy to finally be in the open air. He enjoyed the sun on his face and the sound of human traffic. Being interrogated for hours had not been fun, and he was not planning on ever doing it again.

David suddenly had a curious thought.

Did my clients feel the same way when they talked with me? he wondered. He couldn't understand why he was thinking about his past life until he realized that he had been recounting nothing but history for hours.

All this to keep one police lieutenant distracted? Was it worth it? Will Alex get what he needs to bargain for our lives? David obsessed.

Once they were both a few blocks away from the car that Becky had left for them, they picked up the pace. David took the moments walking to get an update. With little emotion,

Samantha casually brought David up to speed and finally told him that they were supposed to rendezvous at rally point delta, meet with Burns, make the transfers, and get out of town.

David's blood ran cold when Samantha reported in a near-clinical terms how she had to make a decision about whether to blow the car up or not when the family was right next to it. Someone getting hurt or worse was his greatest fear. While David anxiously waited for

Samantha to tell him that she did not kill them, he felt disappointed that he doubted her moral character. David knew that she had done things in her life that crossed the line. He knew that if it was him or Becky, there would be no question they would back down. David was sure that even Burns would have more likely found an alternative.

But Samantha?

Still though, David had never known her to deliberately hurt someone to meet her own needs.

And yet, love can make you do desperate things, he thought.

But if the family didn't move and if she had to choose between strangers and losing the people she loved, would she let them live? David knew that Samantha loved Becky and Emma. David was sure she cared for him as, say, a father.

Does she love Alex Burns? That much to kill? Would it be for him or Becky? Does it matter? he wondered.

Once she finished with her report, David took the opportunity to ask her what he knew might appear to Samantha as an unusually timed question.

"Do you love him?"

David felt Samantha's pace lose a step. Her arm pulled

him a bit tighter, and he felt her turn to look at him. Based on the body language, David had his answer.

"Jesus, David … you've got this thing with bad timing," she retorted.

"It may be months or years before we see each other again, and I want to know you're happy or will at least be safe with him."

It was a long moment, but Samantha answered his question.

"Yes, I do love him. At least I think I love him. As much as I can love anyone, I guess."

Wow…that's really telling.

David knew these emotions had to be confusing for her, but he was sure she did feel things for Alex she had never felt for any man. It seemed obvious that she cared about him.

"You and Alex are going to meet up and be together, right?" David asked.

"Yes. We will go our separate ways for about a month and then will coordinate a plan back to each other."

"Thank God. Your sister was going to kill me if I didn't find out," David confessed.

Over time, David had formed a mental picture of Becky and Emma. Emma was easy; he remembered holding her as a baby. Becky was more difficult to imagine, but there was one thing David knew about her: she wanted him to find out what the deal was between her sister and "Burns" as she always called him.

I guess I can report that Samantha's in love. I hope Alex is aware of this.

Finally, they were in the car, and Samantha asked if David had the keys.

What!? I don't have the keys! I've been in a small room with bad coffee and a headache reciting my story to the police who know what we're doing. Why would I have this car's keys?

David felt his face heat up fearing that he was supposed to have the car keys until he realized that Samantha was laughing as she started the car and drove off.

Now that was just mean. She's gotta be stressed to do something like that. It's something she'd do to Becky, he thought.

David sat back in the seat and finally relaxed enough to smile. It had been a long day. The sun really did feel good on his face. He still felt sorry for misleading the lieutenant.

He seemed to be a nice enough guy, David thought.

Davis pulled her car into the back of the private bank and backed it in not far from a parked ambulance that was positioned in a similar fashion.

That's a good idea to get out fast but how do you get your patient in the back, she wondered as she took a quick look at the plates.

Making sure her car was in park, and her weapons and bag were ready, she refocused on her objective, happy to be well within her own window of forty minutes. This would allow her to be online and in the game in ten. She was also happy she had made the effort to come in early for her shift.

And I got to see a side of Cratty I never knew existed. Granted, it took a national crisis to make it happen. I hope she doesn't get screwed for this breach. It'd be just like Webber to blame her for it. And where the hell is he? He's such an asshole.

As she exited the car, she was vigilant to her surroundings as she pulled out the backpack and locked the car door.

She then made sure to keep one hand on the backpack strap and the other on the handle of her weapon as she scanned all potential ambush points.

Then she heard gun shots.

Damn it! A robbery? Now?

Davis heard a male voice yelling something, and then there were more shots. Instinctively, she extracted her gun from her holster as she lowered her center of gravity to the ground so as to be less of a target. Suddenly, there was a mob of people exiting the bank and headed right towards her until they all stopped in their tracks after seeing she was armed. There were a number of statements like "oh no" and "she's got a gun."

In clear terms, Davis told them to move. Moving quickly from the middle of the parking lot, she found herself just inside the lobby as the door slowly closed behind her. That's when she clearly heard another language demanding something from above her where the vault was, followed by two more shots.

Silence fell until she heard it again: *"Unkraut nicht vergeht!"*

As Davis rounded the corner, weapon drawn and leveled, she started to climb the stars. When she was close to the top of the landing, she saw someone crouched in front of her. From behind, she could easily see and identify the telltale operations center's colored shirt of the guard and knew enough not to shoot one of her own.

Davis could tell that the guard was clearly startled by her arrival.

"Wow! Same side," she said as she held her gun up high, not pointing it at him.

Acknowledging her quickly, he returned to hiding behind a small partition. He looked as if he was covered with sweat. Though the guard was not scared, it was clear he was angry as a result of something going on upstairs and Davis's sudden arrival.

What's up with your hands and head? Scars? Burns or something, she wondered as she looked closer at him.

"Who the hell are you?" the guard demanded.

Davis lowered her gun back down, dropped the backpack on the floor, and took up a similar defensive position behind him.

"I need to get in the vault. I'm with the operations center."

With his gun and attention pointed towards the upstairs landing obscured by a small wall, he seemed more edgy than she liked.

"Where's backup?"

"The operations center is compromised, and I need to get into the auxiliary control room," Davis said as she stood right behind the guard.

"That might be a problem with 'Mr. German' unless I hit him worse than I thought," the guard explained.

"What the hell is a German doing shooting at us?" Davis asked as she replayed the last few moments in her head.

This guy is pretty jumpy. And why did he assume I'm really on his team? she thought.

He looked at her as spoke but kept his gun pointed up the stairs.

"I don't know! Why is the world going into the shitter? It's the Apocalypse," he said as he turned his attention back upstairs where the shooter and the vault were.

Davis had to calm this guy down: *he's ready to shoot anybody at this point.*

"So that was German I heard. Do you know what he said?"

Davis watched the guard wipe sweat from his brow, and he seemed to forcibly slow himself down so that he could clearly think about a potential answer.

Definitely old burn scars. And he's sweating all over the place.

"Something about 'you'll need more to finish me' or something like that. Look! My buddy is up there. I know I hit the bad guy. It's been quiet," the guard finished.

Then in tactical hand signals, the guard indicated that they would need to stack up one right behind the other and rush the shooter. The guard indicated he would take the lead, low and center to the left. Davis shook her head "no" and indicated that she would go first, taking low and center toward the left of the vault. Before the guard could object, she got in front of him and continued to reinforce that he needed to go high and center to the right of the vault.

Assuming that they both knew where the vault door was, she knew that they would capture the entire kill zone in seconds. But the main reasons she wanted take the lead and go low and to the left was because she did not want this guard shooting the bad guy, who might be very important for intelligence. Davis also wanted to cover the vault door, which was more to the left side of the room.

It was easy to see that the guard reluctantly nodded "yes" while Davis gave the signal to move on a count of three. The two readied themselves, their backs against the wall, and then they both bolted up the remaining three stairs and turned the sharp corner. The guard was right on top of her, but it was manageable.

Jesus. A little space, please!

The first thing Davis saw when she broke cover was an unconscious guard sitting in the chair, handcuffed to the side rail on the wall. The alleged German shooter was also unconscious and lying on his back, his hand also handcuffed to the wall rail. The thing that momentarily confused Davis was the fact that the alleged shooter was missing his shirt.

What? What the hell is this? Where's the bad guy?

A fraction of a second may have elapsed before it became clear to Davis that there was a problem. Before she could turn her gun toward the fake guard, a sharp kick hit her hand and knocked the gun clean from her grip. The pain was sharp as he had the advantage of a strong kick while standing above her. But his position also worked against him as Davis's legs uncoiled from her low position upward like a spring at her new target. She hit her target with great force, and she was positive she heard his ribs crack with a sharp exhale.

Good! That's for bullshitting me!

Once she was standing up, she made sure to strip the target's hand of the gun. While successful, she had dislodged a stun gun of sorts.

What? Where's your gun? What…

As she refocused, she felt one elbow graze her chin. Any closer and it would have knocked her out. It was the stomping of her foot and punch in the stomach that separated them from a clinch. The target attempted a punch to her jaw; however, Davis blocked the blow, and then she captured the arm. Once captured, she began her maneuver. Davis thought another rather than having a set on location already, it would have been so much easier to take out one person rather than creating a national crisis. Talk about overkill. What was I thinking? It would have been so much simpler if I'd just paid more attention. Stupid, stupid, stupid," Burns chided himself.

Clearly, I was more focused on Sam than my work, he thought.

Instead of expressing more anger, he smiled. He couldn't help but think she was such a wonderful distraction. He was also surprised that he had no memory from his past life of being distracted by someone he cared about.

Screw it. I'm okay with that, he thought.

He then noticed that his scalp and arms seemed to stop itching.

Hmm. That's odd. Great but odd.

Burns brought the bag up to the desk, took the contents out of backpack and placed all the pieces in his own paramedic bag. Next he emptied the cash, passports, and birth certificates from the cabinet into the same bag. Once he was packed, he lifted the bag and felt his ribs begin to ache. Taking one last look around the immediate vicinity, Burns settled on the unconscious woman before exiting.

"You know, Cougar? You're pretty good. I hope we never meet again. I think you're the type that holds grudges," he said as he slowly walked to the stairs. Suppressing the pain, he focused on taking the stairs two at a time and making a quick walk to the ambulance.

Good thing I'm not being chased.

The walk to the ambulance seemed so much farther than it had when he had arrived. Once behind the wheel, he carefully drove out of the parking lot, barely missing a car speeding recklessly into the bank's parking lot he was exiting.

Readjusting his rearview mirror, he saw a sole woman pulling in the driveway with a vengeance.

"Looks like she's in a hurry," he said to himself as he sped up to put more distance between him, the new arrival and the vault without drawing attention. Then he changed his mind.

"This is an ambulance. An ambulance driving fast won't draw a lot of attention. Not on a day like today. Wake up, Burns."

He was anxious to catch up to the team as rapidly as possible so that they could split up the cache of money and data, and be on their separate ways. It was only a matter of time before all of their diversions were all contained.

No time to waste.

By the time Burns arrived at the rally point, he was very happy to see that Becky and Emma were already at a rest area, two exits into New Hampshire. Just as he was putting his vehicle into park, Samantha and David were pulling into the same area.

Thank God! I'm so glad you both are here, he thought as he tried to contain his sheer happiness. He did allow a broad smile.

Initially, Burns was fearful seeing only Becky, worried that something went wrong, and Samantha and David were caught, or worse. It took only a moment for him to realize that traffic leaving Massachusetts was bad. Moving but slowly.

It was a great understatement to say that Burns was relieved to see that everyone made it. He was very happy to see Samantha especially.

Five years. From strangers to family. Unbelievable.

Upon exiting the car, Samantha immediately hugged her sister, Emma and then him.

Feeling her hug and smelling he real hair, he couldn't help but wonder if he had ever had such an emotional experience before.

Forget about Burns. Just hug back.

Burns could easily see that she was just happy to see everyone again. Feeling her release him, Burns watched Emma and Becky embrace David. Then Emma let David know about some donuts, and that "we went on two runs!"

As David and Emma talked about their days, Samantha and Becky began to look over the luggage and exchange specific pieces. Finally, Burns went to the back of his small SUV and started pulling out some luggage and important items. One of the benefits of having years to prepare for an operation is having time to make sure all rally points and secondary locations have caches of weapons, money, and vehicles. In this particular situation, Burns had been able to abandon his ambulance at the last exit before he entered New Hampshire and exchange it for a less visible means of transport.

I don't know who that was that nearly hit me, but if they're looking for me, they'll only find the ambulance.

Before Burns got down to business, he pretended to steal Emma's nose and eat it. Emma in return jumped so that he could catch her. Burns did but not without a grimace of pain on his face. He was sure his face expressed volumes.

Oh yeah...broken ribs. Probably three.

Once Emma was promptly given back her nose, Samantha was the first to state the obvious.

"Looks like you got fucked up a bit."

"You have no idea."

"Don't tell me it was Cougar!"

Nodding, Burns saw both concern and relief that he was there in her eyes.

"Jesus," Samantha said as she shook her head.

I'm just thrilled we're all alive and together, he genuinely thought.

"Okay, everyone ... children are around," David said as he fumbled to put Emma back in her car seat.

"No problem...we're good to go," Burns said with confidence.

"Yeah ... right."

It was easy for him to see that Samantha did not believe him, but she knew enough not to press now, though maybe later.

He handed Becky and Samantha the hard files of classified protocols as well as the majority of the cash.

Taking a brief moment to look at everyone just to prove to himself they were all there, Burns prepared for his final in person briefing.

"Here's the deal. Continue with the plan to your designated points. Becky and David, I want you to advance your plan by three months and head to your final destination. I have additional passports, birth certificates, and cash to facilitate that. I plan to keep most of the hard drives. If they are bugged and can be located, it will be easy for me to move she had him, and she was preparing to strike her assailant in the head; however, he then countered by turning his whole body into her. Her enemy now had the momentum to push her over the desk with him on top of her.

Damn it! You're not getting away! I bet you're the asshole responsible for all this shit!

The target rolled off relatively quickly. She could see that he was injured but not out. Davis moved a bit faster, and pressed the advantage to kick the target in the ribs. Somehow, the target moved his trunk out of the way and then captured Davis's heel and continued the arc of her kick upward above her head.

Son of a...

This made her fall forcefully backward and hit the floor with great force. Though she was dazed, she could move but only very slowly. Davis knew that most hand-to-hand combat lasted under a minute. She wished she had more time. As she attempted to roll onto her side and get to her knees, she saw her adversary take out a Taser.

Shit! He's taking no chances, she thought.

The stun gun she had knocked out of his hand before was for close-quarter combat. But the Taser allowed for distance. Davis couldn't tell if she was just dazed or genuinely confused, but she kept focusing on why her target kept his gun holstered.

Why won't he use it?

That was the last cogent question she had before the edges of her vision seemed to tunnel in. She then saw the Taser shoot and the electrodes hit her chest.

Damn it...

Davis had been positive she would have passed out anyway from the hit to the back of her head, but now she was really going out. Her eyes continued to dim, and then she felt every cell in her body burn. There was a sudden bright aura she saw as her muscles became rigid. Then there was darkness and a release of tension. If her body was convulsing, she was not aware of it.

Burns was relieved as he watched his adversary finally fall to the ground. For whatever reason, as important as this mission was, he could not bring himself to shoot an unarmed person, no matter how dangerous she was.

She was just doing her job, he thought.

David had warned him that in times of stress and danger, it was very likely that he would have strange thoughts. A few years ago, Burns wouldn't have thought twice about killing someone to complete the mission; it would be no problem for him to be in a firefight, kill as many targets as possible, and then have a beer, eat, and have sex with anyone available an hour or two after his debriefing.

Of late though, for the last several years, he felt remorse and regrets about all the collateral damage he had caused.

No more. No more innocent people.

Burns was struck by the fact that he thought of Matthew's biblical verse:

"For what is a man profited, if he shall gain the whole world, and lose his own soul."

It took Burns seconds to regroup, think of his friends and the next steps in the plan. As he breathed again, he felt the pain in his ribs.

"Maybe it's only two ribs but feels more like three are broken...No matter. Gotta move," he said as he gingerly felt his side as he moved to his unconscious adversary.

Tuning the woman over, he went through her pockets, and found a magnetic key. He approached the door of the vault and ran the key through the slide. Instead of opening, however, a recessed hand-print pad emerged. That was a surprise.

"I guess things do change," Burns said to himself.

Standing back, looking at the door and then the woman, he made an educated guess as to whose handprint he needed. Burns turned and dragged the woman over to the pad, picked her up from behind, and put her right hand on the pad. The pad turned from red to green, and the door hissed open.

Burns carefully placed the woman back on the ground and handcuffed her wrist to the wall rail like the others.

For nearly five years, Burns had waited for this moment. He entered and found a deceptively large room cramped with monitors on a wall. Burns went to look for the files and hard drives, but there were none to be found.

"OK," he said simply as he worked clockwise around the room as his scars burned.

Finding a large cabinet that he opened with the magnetic key, he did find two large cases of cash, all small denominations and most likely unmarked as they would be used to pay off informants, operatives, and hired hands. In addition to the cash, there was one large legal binder with about ten blank passports and a handful of blank birth certificates.

"OK. All good but not what I want," he said calmly.

While clearly the site was a way-station for providing

key resources to field agents and their private, independent contractors, it was not providing the means for leverage he was looking for. Farther in the back of the cabinet, he found files regarding protocols but still no classified information. All of these were helpful but not the goal of five years of work. Burns felt his scalp and hands starting to itch near his scars with greater intensity than before. A sense of panic was forming in his gut.

Burns stopped his searching and then searched counter-clockwise as if that might help him see something different. Oddly enough, it did help him focus on four pairs of monitors and four individual CPUs with four corresponding external hard drive docking ports.

"Empty. All empty. If they're not here...where would they be..." Burns stood still in the middle of the small room, his hands rubbing his head as pain radiated from his lungs every time he breathed.

"I'm missing something. Something simple," he said to himself as his scalp and arms continued to itch.

Then it hit him. He had watched various operations center staff enter the bank always with a briefcase or bag.

"Wait. She had a bag. Black backpack when she had come in."

Burns ran to the stairwell, saw the bag, and opened it.

As he sighed, he ignored the pain from his broken ribs as he finally saw the external hard drives he had planned on seizing for five long years.

"Five years of black-op secrets for five lives...I can't believe they changed the protocol! If I'd known that they *physically transported* the hard drives from one secured position to another rather than having a set on location already, it would have been so much easier to take out one person rather than creating a national crisis. Talk about overkill. What was I thinking? It would have been so much simpler if I'd just paid more attention. Stupid, stupid, stupid," Burns chided himself.

Clearly, I was more focused on Sam than my work, he thought.

Instead of expressing more anger, he smiled. He couldn't help but think she was such a wonderful distraction. He was also surprised that he had no memory from his past life of being distracted by someone he cared about.

Screw it. I'm okay with that, he thought.

He then noticed that his scalp and arms seemed to stop itching.

Hmm. That's odd. Great but odd.

Burns brought the bag up to the desk, took the contents out of backpack and placed all the pieces in his own paramedic bag. Next he emptied the cash, passports, and birth certificates from the cabinet into the same bag. Once he was packed, he lifted the bag and felt his ribs begin to ache. Taking one last look around the immediate vicinity, Burns settled on the unconscious woman before exiting.

"You know, Cougar? You're pretty good. I hope we never meet again. I think you're the type that holds grudges," he said as he slowly walked to the stairs. Suppressing the pain, he focused on taking the stairs two at a time and making a quick walk to the ambulance.

Good thing I'm not being chased.

The walk to the ambulance seemed so much farther than it had when he had arrived. Once behind the wheel, he carefully drove out of the parking lot, barely missing a car speeding recklessly into the bank's parking lot he was exiting.

Readjusting his rearview mirror, he saw a sole woman pulling in the driveway with a vengeance.

"Looks like she's in a hurry," he said to himself as he sped up to put more distance between him, the new arrival and the vault without drawing attention. Then he changed his mind.

"This is an ambulance. An ambulance driving fast won't draw a lot of attention. Not on a day like today. Wake up, Burns."

He was anxious to catch up to the team as rapidly as possible so that they could split up the cache of money and data, and be on their separate ways. It was only a matter of time before all of their diversions were all contained.

No time to waste.

By the time Burns arrived at the rally point, he was very happy to see that Becky and Emma were already at a rest area, two exits into New Hampshire. Just as he was putting his vehicle into park, Samantha and David were pulling into the same area.

Thank God! I'm so glad you both are here, he thought as he tried to contain his sheer happiness. He did allow a broad smile.

Initially, Burns was fearful seeing only Becky, worried that something went wrong, and Samantha and David were caught, or worse. It took only a moment for him to realize that traffic leaving Massachusetts was bad. Moving but slowly.

It was a great understatement to say that Burns was relieved to see that everyone made it. He was very happy to see Samantha especially.

Five years. From strangers to family. Unbelievable.

Upon exiting the car, Samantha immediately hugged her sister, Emma and then him.

Feeling her hug and smelling he real hair, he couldn't help but wonder if he had ever had such an emotional experience before.

Forget about Burns. Just hug back.

Burns could easily see that she was just happy to see everyone again. Feeling her release him, Burns watched Emma and Becky embrace David. Then Emma let David know about some donuts, and that "we went on two runs!"

As David and Emma talked about their days, Samantha and Becky began to look over the luggage and exchange specific pieces. Finally, Burns went to the back of his small SUV and started pulling out some luggage and important items. One of the benefits of having years to prepare for an operation is having time to make sure all rally points and secondary locations have caches of weapons, money, and vehicles. In this particular situation, Burns had been able to abandon his ambulance at the last exit before he entered New Hampshire and exchange it for a less visible means of transport.

I don't know who that was that nearly hit me, but if they're looking for me, they'll only find the ambulance.

Before Burns got down to business, he pretended to steal Emma's nose and eat it. Emma in return jumped so that he could catch her. Burns did but not without a grimace of pain on his face. He was sure his face expressed volumes.

Oh yeah...broken ribs. Probably three.

Once Emma was promptly given back her nose, Samantha was the first to state the obvious.

"Looks like you got fucked up a bit."

"You have no idea."

"Don't tell me it was Cougar!"

Nodding, Burns saw both concern and relief that he was there in her eyes.

"Jesus," Samantha said as she shook her head.

I'm just thrilled we're all alive and together, he genuinely thought.

"Okay, everyone ... children are around," David said as he fumbled to put Emma back in her car seat.

"No problem...we're good to go," Burns said with confidence.

"Yeah ... right."

It was easy for him to see that Samantha did not believe him, but she knew enough not to press now, though maybe later.

He handed Becky and Samantha the hard files of classified protocols as well as the majority of the cash.

Taking a brief moment to look at everyone just to prove to himself they were all there, Burns prepared for his final in person briefing.

"Here's the deal. Continue with the plan to your designated points. Becky and David, I want you to advance your plan by three months and head to your final destination. I have additional passports, birth certificates, and cash to facilitate that. I plan to keep most of the hard drives. If they are bugged and can be located, it will be easy for me to move faster. Sam ... take this hard drive as my insurance. If I miss contact with you by twenty-four hours, let Becky and David know. If forty-eight hours pass without contact from me, run silent, regroup, and start the campaign of letting the media access the hard copies and hard drives. Hard copies first in pieces and then flood the Internet."

"'And they shall not leave in thee one stone upon another,'" David uttered after he buckled Emma in her car seat.

Burns smiled. He could always count on David for a literary and biblical reference that fit the occasion.

"Luke 19:44. By the way, David, I had more of those thoughts when I was nearly killed back there," Burns commented.

"Kind of makes sense, Alex. I assume that when you're not working, planning, and reconfiguring, you're praying and meditating in one form or another. Are you thinking you'll eventually want to become a Catholic or more Orthodox? Maybe Jewish?" David suggested. Burns smiled. He was sure he would miss David greatly. The guy who saved him so many years ago was still trying to help him find answers.

He gets me, Burns thought.

"I think Catholic, though I prefer their Latin High Mass," he answered thoughtfully.

David was shaking his head in approval.

"Fine choice ... much shorter services than the Orthodox liturgy while capturing the key points. Latin would appeal to your sense of tradition. Good choice, Alex."

Burns continued to smile and then turned to address the whole group.

"Well, that's it, team. If all goes well, we'll be catching up in real time in twelve months," he concluded. As a wave of sadness passed through him, he knew that unlike past missions from years ago, this team had become something more. It was his family, a family he might not see for a long time.

That's more than a good enough reason to be sad, he thought.

"Really? You two aren't going to see each other for twelve months? Please," Becky challenged.

He could see Becky peering at Samantha.

"All right. I plan to join Sam much sooner than that," he confessed.

As Burns watched Samantha's smile, he was happy to think that he might be the reason. With that, he watched her hug her sister and kiss Emma good-bye. After a few moments he observed the ritual he was familiar with as he watched how Samantha said good-bye to David. It was a unique relationship that he marveled at. Their version of hugging had always consisted of a firm handshake and the usual scripted sayings.

"Touchy-feely" was just not part of their relationship. Burns found it ironic that the psychologist and the nurse were the least comfortable with public displays of emotion.

It's just not their relationship, he reminded himself.

"Ms. Littleton … take care of yourself."

"Dr. Caulfield, always a pleasure, and until next time," Samantha said.

"You know where to find me," David responded.

"Sure do," Samantha said sweetly.

However, Burns noticed he did make a change when he addressed him.

"Alex? Take care of her, please," David smiled.

Burns turned away and then looked back at David, who was looking away too and still smiling.

Burns looked at Samantha and gave her a look as if to say "something more is going on here." She quickly picked up on the nonverbal communication and looked at Becky in a curious way. Burns knew that Samantha had been hounding Becky about her and David's relationship.

"So are you and David going to stay together? Don't you think Emma needs a mother and a father? He kind of is already," Samantha asked quietly.

Becky smiled as she entered her minivan as if she didn't hear Samantha's question.

"Call me later, gator," was her only response.

As the minivan pulled away, Emma could be heard yelling from the back, "I love you!"

Burns wondered if even Becky knew the answer. Suddenly, Burns realized he was alone with Samantha. He was at a loss for words. She found the words faster than he did.

She's always faster than me at this.

"Alex, I am going to head out, but I plan on seeing you in two weeks."

Burns was about to protest, but she waved him off.

"In two weeks, this plan will either work, or it won't. Either way, I want to be with you, so don't argue with me, and we will get along just fine."

No. I'm not going to fight this. Not this time, he thought.

And that was it. Burns had to decide if he was going to argue and hold ground or yield.

Well, "no" is not even a real option.

Burns had only one response: "Acknowledged."

With a quick hug and a kiss, Samantha got in her car and drove to her off-site storage facility, where he was sure she would undergo a complete change and transform herself yet again.

"I wonder what wig she'll wear this time?"

Burns was alone, but unlike many other missions, he was remarkably happy and hopeful.

But there was still some work to be done.

Chapter 18

"Fama crescit eundo"
"The rumour grows as it goes,"

– Vergil

Present Day – May 2

ONCE THE EMERGENCY ELECTRICITY went out, there was no electricity to keep the locks for the cell doors secured at the North Reading Police Department. As a result, Andersen had been outside for an hour, corralling the detainees and criminals. While moving the citizens was a priority, making sure the real bad guys did not get away was a matter of public safety.

Wow! It's a different world from early this morning, Andersen noticed when he emerged from the interrogation room into the light of day.

Sirens could be heard everywhere, heavy traffic on the highway could be heard from his vantage point, and there was a sense of confusion as seemingly all computers on the planet were going haywire at the same time.

"So this is the way the world ends. Not with a whimper but an old-fashioned dial tone," he said to one of the cadets.

The cadet smiled but Andersen was sure the young man had no idea what any of what he said meant.

These kids should read more.

Once there was some semblance of order, Andersen returned to the interrogation room with Jefferies and found it empty. Stepping into the empty room, he looked on the desk and saw that his notes, along with his material witness and possible terrorist, were gone.

This is not real. This can't be for real.

Andersen stood in the room for about a minute, taking in the full gravity of how he had been duped and allowed a key player to walk right out of custody.

That bastard. That brilliant bastard...

By nature, Andersen was not a violent man. He was not prone to outbursts at all. It just wasn't who he was as a person. Not as a father, cop, soldier, or famed interrogator. As Andersen was about to walk out, he turned suddenly on the desk and flipped it over with a guttural yell.

"God damn you Caulfield! Shit!"

Andersen stood for a moment longer in the silent room, his hands pressing into his head through thinning hair. After a moment of eerie silence, he found himself impressed with Jefferies's ability of remaining invisible and allowing his boss to have a moment of frustration.

Andersen regrouped and issued an order: "Search everywhere and find them."

"Will do, boss," Jefferies responded.

Good answer! You should make captain by the end of the day!

At least someone knows how to do their damn job, Andersen thought as images of Caulfield and the cadet walking out of his police station blinded him with rage. It was a foregone conclusion that the cadet had to be one of the Littleton sisters.

It was probably Samantha. And I bet the other woman with the child was Becky and Emma...they just walked in and walked out of here like they owned the place!

"They walked into my house, messed with the place, and then just walked Caulfield out. Son of a bitch! That takes pretty big balls," Andersen muttered to himself as he marched upstairs to find his gun.

Pretty big balls. And to sit there and tell me the truth. Yeah...I bet it was all true. And then just walk away – what the hell... Why would he do that?

After twenty minutes, Jefferies brought Andersen up to speed about the search and told him that it was Officer Dempsey who had found the red wig, cadet jacket, and Caulfield's suit jacket and tie just outside of the emergency exit. Andersen was about to ask why the alarm had not been triggered, but then he remembered the electricity and backup generator had been off-line. Andersen shook his head. It had all been part of their plan.

This is just crazy. They planned everything. Everything right down to diversion and extraction. But why? Why here? Why me? What were they after? Was it all true? Burns too?

Sitting with his police windbreaker on and armed with his semiautomatic and three clips of bullets, Andersen decided enough guessing and to get back in the game.

"Jefferies! Find our FBI contact and then get me his

boss. I want everyone to find Caulfield and the woman who walked right in and out of my police department without a problem. Find them now!"

It was only a few minutes later when Jefferies came back with a name and a location for the FBI deputy director.

"His name is Helms," he said quietly.

Dempsey surprised him when he somehow managed to get Helms on the CB radio. Andersen informed Helms that he was now missing a possible link to a terrorist cell. As luck would have it, Helms was in Andover, about ten minutes away. They agreed to meet halfway.

Well, maybe the tide is turning.

With no computers and half the digital landlines and cable off-line, that left only two-way scanners, CB radios, hardlines and walkie-talkies, and no rapid means to distribute a be-on-the-lookout notice for David Caulfield and Samantha Littleton.

Andersen knew that with all the chaos going on and no Internet access, texts, and activity on smart phones, it would be hours before the notice made it out to the field in the Merrimack Valley.

"This is just great! 1970s technology is state-of-the-art again with all the computers down! What's next? Are we going to have to find a mimeograph machine to make posters?" Andersen complained to Dempsey as they drove.

"Boss? This mimeograph machine? If it needs a CPU, it'll probably not work. All the computers are messed up bad," he said in earnest.

Andersen turned to look at him for a moment and simply let it go.

Just great.

Andersen thought about all of this when he arrived at the designated location. It was easy to see who was in charge.

Yup. He's the older-looking guy like me, he thought mournfully.

He noticed that Helms was flanked by two of the most computer-savvy agents he had ever seen.

Jesus...how old are these kids? Is the FBI recruiting right from high school?

Once he met face-to-face with Helms, Andersen summarized the entire interview, the deception, and the target. Andersen thought one of Helms's computer gals, a young woman named Janeson, was attempting to establish contact with some federal agency called the operations center. It took a while, but she was able to get a manager on a working phone, someone named Denise Cratty.

Helms took the phone from Janeson and started to listen and then interrupted.

"Wait a minute ... I have the detective that interviewed our lead witness right here. Let me put you on speaker phone. There are two of my agents and him."

Helms pressed a button, and everyone intuitively huddled around the cell phone.

"Everything I am about to tell you is classified and needs to be treated as such. An hour and a half ago, my operations center was compromised by a crippling cyber-attack. Fortunately, with your house being hit first, it gave us a jump to shut things down and get a backup plan going. We had four real-time missions going, so we relocated our operations center to Hanscom Air Force Base. In the meantime, one of our senior-ranking managers, Jill Davis, was deployed to our auxiliary control room to reestablish and continue all operations until we were fully operational again."

Andersen noticed that Helms had clearly recognized Jill Davis's name.

"Davis said she would have the operation going in less than sixty minutes, and that was fifteen minutes ago, and it's still dark. She's not responding to any modes of communication, and if she is not responding, that means something bad has either happened or is going on."

Andersen looked right at Helms. He shook his head, positive that Cratty's assessment was accurate.

"Gentlemen, I need all available resources to block off all entrances and exits at that location. We are en route and ten minutes out. We need to secure the area, but I really could use your help to make sure nothing gets out. Local PD is on the way and will be there in two minutes. Will you help?"

Andersen could tell that this manager, Cratty, was a capable leader and a thinker. He could also tell that she could make second fiddle sound important to anyone. Regardless, Andersen wanted answers, and it was easy to see that Helms seemed both pissed and worried.

I thoroughly get that, he thought.

Helms looked at Andersen, and it was his turn to give a positive nod. They all marched in unison to a large SUV.

"Give us the address, and we'll have both our teams in place," Helms requested.

As he sat in the back with Helms, Andersen saw a box of cell phones that two young men were looking through. After the vehicle launched, Andersen joined in to sift through the box to find one that worked so that he could round up any resources left from his beleaguered town.

At first, Davis felt weightless, very light, as if floating above the clouds at first before she found herself diving through a cloud bank. Breaking through, she could tell she was flying...she was flying high above the ocean. In the distance, she could see a ship, a dark ship, as the wind and the salt air filled her lungs, and she felt thrilled to be flying again. Even at her great height, she could feel the mist from the ocean spray her body. But then darkness suddenly fell as the wind died off and salt air faded as if a great storm was breaking. Looking above her, she saw an eagle. It was an enormous eagle blocking out the sun's rays.

Look at the size of that monster! And...what are they doing? Attacking the eagle? That's crazy!

Even as the eagle flew lazily above her, almost touching the sun, five falcons flew in formation as if to attack the great raptor. Small, cunning, fearless, the falcons flew swiftly forward as darkness enveloped both them, the sky, ship, and ocean. Watching the darkness fall, she felt a rush of anger as the opaque expanse marched towards her. Seeing no place to flee, she looked in the direction of the falcons and decided to join them.

It's better to die well...

Flying as fast as she could with the wind at her back, she felt at peace as she joined the others in formation. But there was something she was hearing...words...

That's strange...What? What was that? Did someone say something?

As if coming out of a deep sleep, she heard a woman's voice saying something in the distance:

"Auxiliary control room is breached. Davis and guards are alive but down...All materials in the vault, external drives included, are gone..."

Denise...Cratty is that you?

As she tried to listen, she felt herself slipping away until she felt a pressing sensation on her arm. Then she heard more noise - maybe someone calling her name. Feeling as if she could no longer breathe, she nearly jumped out of her skin as her fists balled and her eyes snapped open. She was about to strike but her right arm seemed dead, and her left felt like it was tied down by weights.

"What the hell! Where is he?" she called out.

It took her a moment to realize that Cratty was trying to wake her up.

"Stand down, Davis. It's me," Cratty responded as she firmly griped her aching shoulders.

After a moment, Davis figured out where she was.

Vault, auxiliary control room, attacked...stunned...

"Are you all right? What happened?"

"I was tricked and attacked. Single, white man in his late thirties. Professional, maybe former special forces. Knows his hand-to-hand and knows tactics. Scars on his hands, arms, head...like he was burned or something...Why are you here?" The haziness was starting to break up.

"The missing field agent, Anthony Maxwell, was at the site of the shooting in North Reading. He's a casualty," Cratty explained.

Davis looked directly into Cratty's eyes, and she could

tell that Cratty had a mix of emotions: *sadness and raw anger.*

Damn it...Maxwell? He's pretty high up on the food chain. Who the hell could get the drop on him? Davis thought.

Hearing that someone was dead reminded Davis of her parents. Instinctively she went to her throat with her free hand to see if her mother's necklace was still there. Davis was relieved it was not lost in the fight.

"How long have I been out?" Davis jumped in again.

"No idea. When you were not up and running after sixty minutes, Chairman Daniels sent me here, right after another agent confirmed you were down. She's after your guy, though Daniels is not telling me who the target is and who's on his trail now," Cratty said as she rubbed her own temples.

Secretive? Now? Davis thought as she felt her body ache.

"Why is Daniels involved? Where's Webber?" Davis asked.

"Director Webber is still on vacation and out of contact," Cratty said with a tone.

Jesus. Not much love there.

Davis's body seemed to jolt with recognition as if she forgot something very important as she interrupted Cratty in mid sentence.

"Where's the bag?!" Davis's eyes began to look everywhere.

Cratty shook her head as she unlocked Davis's handcuff and freed her hand.

"It's gone, Davis. So is the cash, classified protocol files, possibly blank documents...the hard drives, all gone. Only thing he left behind were the weapons."

Davis's mind seemed to go silent.

Am I in shock? She didn't know.

Again, she thought, *no weapons.*

The guy clearly knew how to use his body as a weapon, and yet he did not kill her or the guards.

He had the means, the opportunity, and the skill. Why?

Davis's thoughts were now interrupted by a familiar voice she heard booming in the distance.

"Move aside, young man. I am the FBI deputy director of this region. This act of domestic terrorism and this crime scene should be under my control."

Oh boy. It's Helms. God save us all.

Cratty intervened and allowed him in. Davis didn't know the guy beside him, but she was sure he was important. He had to be as old as Helms.

Regardless of the situation, Davis was glad to see him.

Now on her feet, she could tell that feeling was slowly returning to her arm. In addition, she felt a big bump on her head, a laceration on the side and front of her face, and dry blood in her hair, mouth, and knuckles.

The bruises are going to be really bad in the morning, she thought.

Helms slowed to a stop, looked her up and down, and just took in the sight. Davis anticipated his response.

"Davis … you look like flattened horse shit."

Yes. You are consistent — charming in a marine way, Davis thought.

"Helms … it's been too long," she responded with a very faint smile.

The man next to Helms was next to jump in.

"Does anyone know who Alexander J. Burns is?" he asked innocently.

Davis felt her eyes widen and her breath come up short.

Every operations center person in the room stopped what they were doing and set eyes on Andersen, as if he had invoked Satan's name on an innocent child. The response alone answered the question. It was obvious to Davis that Helms and this guy had no idea who Burns was and were feeling out of the loop. Davis looked at Cratty, who was also still.

"God damn it," Cratty said as she took out her cell phone to see if any signal was restored.

Davis responded next, feeling distracted and angry at the same time.

"Jesus Christ ... that was Alex Burns."

I was so close...I almost had him. Shit!

Davis refocused her thoughts and looked up to see that Helms's friend, Andersen, was obviously frustrated.

"Okay. Can someone bring me up to speed because I had a material witness in my custody who claimed this was all part of a master plan of more than four years."

"Where is your witness now?" Cratty asked urgently.

Davis's hope for a possible break rose like a heated geyser. Then her stomach fell as Andersen shook his head no, killing any hope of salvation.

"Gone. Like he was never there. My notes are gone. My technicians tell me that all audiovisual recordings were fried as a result of an electrical surge and then totally nuked by a vicious virus or worm from within my own police station, which means this bastard had access to everything in the station. And he just walked right through and out like it was the Boston Public Garden!"

It was evident to Davis that Andersen felt stupid, fooled, and idiotic.

Almost as bad as I look to everyone here, she thought as she looked at her bruised knuckles while fiddling with her necklace.

Cratty looked around at her people and was obviously as frustrated as Davis was at the whole situation.

"Stop gawking, people, and start a sweep for all the missing items! Burke, get everyone out to a five-block radius and search home to home if you have to. If any of you want to go home tonight, find Burns and my data. Now, people!"

Davis saw that Alpha shift was not used to their boss barking orders. They jumped into action. Cratty then turned her attention back to Davis. Davis could easily see that there was regret in her eyes.

"I have to report to the chairman regarding the extent of breach, though I'm sure he already knows," Cratty sighed.

"Davis? I'm sorry. Bring our colleagues up to speed about the infamous Mr. Burns."

Davis knew what that meant. Alexander Burns's file was top secret. Classified information was on a need-to-know basis only. By Cratty giving her the green light to bring Helms and Andersen into the loop, it meant that probably both of their careers were either completely destroyed or significantly damaged. Her agency usually had issues with staff that failed to protect national security. At the same time, Davis agreed with Cratty that these two guys needed to know.

Dizzy, Davis leaned up against a wall as another agent

handed her missing gun to her. Davis checked her gun and saw she needed to put the safety on before she started her unofficial briefing.

"Alexander Burns was part of the team that went in to kill Oman Sharif Sudani on May 1, 2011. The plan was straightforward, a real search-and-destroy operation."

Davis noticed that Andersen had a look of recognition on his face.

"Did the intelligence on his location come from a detainee at Guantanamo Bay who directed you to couriers?" Andersen questioned.

Unsettled by his knowledge of classified data, Davis gave the nonverbal for "yes", but she didn't know how he could have possibly known that information.

It was evident by his silence that he planned on not letting her or anyone know how he had that classified data. Now it was Helms who was totally out of the loop.

"Please go on, Davis. At least I am brand new to knowing this guy," Helms said.

"Anyway, the plan was delayed because of a logistics issue that Burns was managing. Even though we had the presidential order to proceed, and Chairman Daniels signed off on the plan, Burns had reservations, fearing that the mission could be jeopardized. As it turns out, he was partially right. A helicopter did crash. Legend has it that Burns got everyone out but something happened and he was KIA."

"But I recall that there was no loss of life in taking out Sudani," Helms contributed.

"True. He wasn't KIA as much as he was MIA for a short period of time. It gets really murky, but somehow, he got injured. He was found by the Red Cross and finally identified, and then he was put into special care at the veterans' hospital stateside. After that, there was an incident where something happened to him. I still don't have full clearance to access Burns's entire file. He was involved in missions I had never heard of or seen any reference to. Burns's record was completely sealed by our boss, the chairman himself. The word in the field though was that Burns changed, and then we lost track of him altogether. A three-man team watching him was killed. All three were seasoned — one agent and two associates. Some believe he is dead too."

Davis slowed down as her headache was really taking form. She continued with her briefing of sorts.

"Many of us are convinced he's still out there, seeking revenge and waiting to strike. Why he's pissed and wants to get back at us is anyone's guess. I sure as shit don't know. Still, more believe he has joined forces with either the North Koreans or Chinese or both. Regardless, everyone at my level knew that the chairman dedicated resources to find Burns and never did." Davis stopped.

Her head was really pounding now. Her trip down memory lane only heightened her anger at being so close to capturing Burns and his escaping with classified data.

Damn it! I almost got him, she angrily thought.

"So it's probable that everything David Caulfield said was true. His wife is killed, Burns saves his life and changes, he trains an entire team to get to this special room, and then they all just disappear. All this to get critical, classified data? What for?" Andersen said, perplexed.

Davis fell silent, as she watched a very angry looking

Cratty approach with two men behind her. It was easy to see that both Andersen and Helms could see anger on Cratty's face.

Well...I guess the phone call with the chairman did not go well. No surprise there. For a pretty woman, she has quite the scowl on her face. I wonder how bad it is, Davis thought.

"Gentlemen, thank you for your assistance, but I need you to leave this crime scene. Ms. Davis, you need to surrender your gun, badge, and ID and return with me to the bunker," Cratty said with no enthusiasm.

At that moment, two relatively large agents came from behind. Not too close, but clearly, they were there to ensure compliance. While Davis was a fighter by nature, she knew this had to run its course. She knew that Helms knew this bureaucratic nonsense was bullshit and that the former marine would help her later on if she needed any assistance. Davis was pretty sure she would need help. Before she fell between the guards, Cratty stepped in her path.

"I'm sorry. It's complete crap," Cratty said in a low voice. She sounded quite genuine and her eyes seemed determined.

I wish I had gotten to know you better, she thought.

Davis sighed and handed everything over and followed Cratty with her two guards in tow.

One possible upside to all of this might be that I will get to meet the chairman in person. I wonder if I still have a shot at that promotion, she thought sarcastically.

He was a man of mystery. Davis didn't care much for the director, though. She had been positive that Cratty was going to get the promotion to deputy director before this debacle.

After? We had Burns and his "new team" and let him go? Both of our careers are done here, she was convinced.

As both men watched Davis, Cratty, and her team leave, Helms gave Andersen the nod to leave as well. Andersen agreed.

Nothing more to learn here, he thought.

As they were both on their way out, Andersen took a detour into the bank. He noticed that the place was not crowded, but there was staff around. He was surprised that Helms didn't even ask him what he was doing.

I bet he had the same idea.

Andersen casually asked for the day manager to hopefully learn some information. Without much of a preamble, Andersen saw that Helms recognized the approach — "active cop/quiet cop," a different and more effective way of obtaining information than the expected, worn-out "good cop/bad cop" approach. Once the bank's day manager arrived and Andersen asked some questions about the camera and the closed circuit TV system, it was apparent that the computer issues had compromised a great deal of the bank's tapes by turning them into black afterimages or unrecognizable pixelated forms.

On a lark, Andersen asked what might have seemed to be an unusual question.

"By the way, you wouldn't happen to know an African-American or Latino man in his late forties, very professional? He also happens to be blind and has various injuries on his face."

The response was immediate as the manager looked

down for a moment as if to recount all possible patrons just to make sure he had the right one.

"Oh, you must mean Mr. Coleridge. He is a very interesting man. He always came in with his assistant, a very attractive, dark brunette woman. She always wore a nursing uniform of sorts," the manager explained.

Yes. Of course he brought her along. How else can you do reconnaissance on a place if you're blind: he's the way in and the distraction, and she's the eyes. Making a disadvantage work for you...that bastard.

Andersen tried to keep his fists from balling up and

maintain a casual, calm tone. Without seemingly skipping a beat, he kept going while Helms watched the magic happen.

"I was wondering when the last time Mr. Coleridge was here. Samuel T. Coleridge. He is associated with an industrial company, correct?"

It was evident that the manager knew that Andersen had detailed information on a bank customer.

"Actually ... I really can't give out that information on a private customer."

Andersen knew that the manager had to offer some resistance and not casually release such confidential information, but truth be told, he was in no mood to be blocked. He just moved ahead.

"I'm not asking for his social security number, though I can get that in ten minutes..." Andersen said as he looked at Helms.

Helms, who was still wearing his FBI windbreaker and tactical vest, clearly picked up on the strategy and opened his cell and started talking:

"Janeson ... get me the social security number, background information, and a judge's order for Samuel T. Coleridge, aka David Caulfield. While you're at it, get the names of all the employees and persons of interest who may bank here. I also want you to see if reputed mobster John Murphy has any holdings here. I am going to need a federal judge and twelve — no, wait a minute... Detective Andersen, a bank this size probably will need about thirty to forty agents to go through all of the files and security boxes, right? Could we borrow some of your guys to help out?"

Taking his time as if thinking about it, Andersen first nodded before he gave a detailed, fictional answer.

"Actually, I can only give you fifteen for now with the national crisis and all going on. But I'm sure I can get Commandant Welch from the state troopers' barracks to help out. About twenty guys there, I bet," he added.

"I don't know how the judge is going to take hearing that we received no cooperation from the bank personnel during this national crisis," Helms said to Andersen with some of the best acting he hadn't seen since his work at Guantanamo.

Nicely done. Obviously this is not your first time doing this, Andersen thought, concealing a smile.

It was evident to Andersen that the manager visibly weighed the needs of his possibly two hundred high-profile clients who valued their privacy against one man's. It had to be an easy decision as long as word did not get out. It had to appear to the manager that if he refused access, the police and federal agents would tear his bank apart, and he would lose those precious patrons.

After he cleared his throat, the manager gave up his stance on client confidentiality.

"Mr. Coleridge owns a green energy company called Renewable Green Energy, Inc. Up until last weekend, the company had some cash in our bank, but mostly, they were auditing and working on making our bank more energy efficient."

"Did Mr. Coleridge ever ask about upstairs?" Andersen asked.

"He actually had an incident where he accidentally went upstairs and ran into the security team up there about three months ago," the manager confessed.

Yup … of course it was accidentally. It's all just one big, awful coincidence. No one would question a vision-impaired person getting lost. This Caulfield guy…his "team" had everything covered. Damn it, Andersen thought to himself.

Andersen sighed and looked at Helms. Closing his phone without saying goodbye, Helms's jaw hardened and eyes narrowed.

As both men walked out in silence, Andersen quietly recapped the entire situation both in and outside interrogation room 8.

So that's it. It's over for now. David Caulfield had been telling me the truth all along. It was all planned out, including access to information and data for this very location, which just so happened to be the former employer of this Alex Burns. This wasn't an inside job! Davis is going to be scapegoated for just doing her job, a victim in the right place at the wrong time. This plan had been in the making for years — times, locations, decoys, and resources - all carefully prepared. I bet there were backup plans to the backups: there were undoubtedly fail-safes and codes in place that were there "just in case." Unbelievable. This was all done by a team of civilians who had been trained by a specialist so that they could compromise the US government. But why? What are they going to do with the shit they stole?

Andersen knew that if this was all true, someone — David or Burns or someone from that team — would be calling him or someone like him about their demands. While he had no idea about what exactly Burns and his crew had walked away with, he had no doubt that Burns would be willing to rain down hell on all of them to protect his team.

What did you get? What's so important to do all of this? What balls!

As both men walked out of the bank in silence, Helms was the first to speak.

"He's going to contact you, you know."

"Yeah, it makes sense. Even though I am unlisted, I bet he will get through to me," Andersen responded.

"Do you want a detail? It would be very easy for me to set that up," Helms offered.

It only took Andersen a moment to assess the risk and dismiss it.

"No. Burns and his team walked in and out of my station, twice, compromised two federal agencies, shut down the entire Northeast, and stole top-secret files. All this before lunch. Anyway...Burns needs me. He won't hurt me. He hasn't hurt anyone really. The early ballistics reports show that your guys killed the guys from Boston, and those guys killed the other federal agent. And in light of the explosions, smoke, and fires, the only real damage has been property, egos, and national security. Burns has gone to great lengths to create havoc and chaos while limiting collateral damage. No ... he won't hurt me, not now anyway. Maybe if I piss him off, but not now."

Andersen no longer felt anger about the deception. He was fascinated by everything this group had done. The level of detail and patience was staggering to imagine.

I really should hate you, Caulfield...

Andersen had to get back to work. He knew that while both of them would proceed with their investigations and continue the search, he was convinced that Burns and his team were long gone by now. They had had at least a 90-minute head start and had been five steps ahead all along.

With the computers still down, they could be eighty miles away and dug in deep. The way this team operated, they might be watching me and Helms right now, Andersen mused.

Probably not, he thought better.

Chapter 19

"Mundus vult decipi, ergo decipiatur"
"The world wants to be deceived, so let it be deceived"

Two Weeks Later – May 16

"THE MERRIMACK VALLEY CRISIS" as it was now called, lasted about four days before it stopped being the lead news. The bombs that were planted never blew up. While they looked scary, they were never rigged to explode, just theater. Weapons of mass destruction were never found, let alone detonated. Helms's team found the computer and master server in a building complex very close to his own police station that had compromised the operations center. In one of his last communications with Davis, Andersen found out that the operations center was now somewhere in one of ten states now, and they now had a plan to have two operational at all times. The auxiliary control room's fail-safe method was discontinued. In light of the breach of classified information, the protocol for relocation and operations had been completely changed to work parallel of each other rather than sequentially.

On a national level, it took nearly fourteen hours to rectify the effects of the computer worm that compromised thirty-two million American computers. Since this was everyone's personal computer, people were really pissed and wanted justice. This kept the story alive on the front-page news for an two additional days. It was apparent that antivirus companies had lost control of their products and the consumers' confidence, and then lost a great deal of money. It was a massive public relations disaster. It was also clear that various private security agencies were ill-prepared for such a cyber-attack.

The irony of all this, Andersen suspected, was that the economic damage and the anxiety of having your computer turn against you was more troubling to the average person than the fact that civilian lives could be lost.

Helms told Andersen in one of their weekly meetings that information had been omitted from the media and kept from the public. Specifically, all active security cameras and surveillance equipment at the time of the crisis had been profoundly compromised. This fact was kept quiet to make sure that criminals and terrorists did not take advantage of the weak infrastructure. In fact, the missing classified data and who might be involved in the incident had also been left out of the news as well.

By day seven, the conspiracy theories started emerging. Andersen thought the best one was that the attacks were a joint Chinese, North Korean, and Middle East "trial attack," and that the "federal, state, and local governments effectively contained and preserved the nation's infrastructure." It eventually became the common mantra following every news report.

It's a good spin. Better than "a bunch of pissed-off citizens and a former spy nuked our entire national security. Have a great day and sleep well," Andersen remembered thinking at the time.

Andersen had a hard time adjusting to the realization that all the mayhem could have been profoundly worse, and yet all of it had been successfully committed by a group of average citizens that were clearly motivated.

But why? What would motivate two sisters and a psychologist with a baby to take on the US government? Why did they do any of this?

After two weeks, Andersen still contemplated all this from his leather chair while he drank a single-malt scotch.

With the television off, and the kids settled into their evening routine, he was enjoying a quiet moment with the plan of reading a magazine so as to get his mind off of Burns's ultimate motive.

Maybe the whole Burns affair is over? No. I wish. Not for me. Not for Helms, either.

Andersen knew he would never rest, even if Burns or David never contacted him. He was pissed that he fell neatly into a plan and was used as proficiently as a surgeon might use a scalpel. He was not a drinker by a long shot; however, he had started taking an interest in malt scotches a few years back, and now seemed like a good time to actually try some. The bottle had been a gift from Helms. He too was still determined to find Burns and his crew because, as he put it, "they screwed up my region."

I guess it was just as personal for him too, he thought as he sniffed his drink and took a sip.

Now why did I stop drinking?

After enjoying the strong warmth of the drink, Andersen settled on looking for a magazine article that might be interesting. That is, until his wife's cell phone kept going off.

Damn it! I can't stand it when she leaves it on. Why can't she turn it off when she leaves it?

With all the cell phones and other communication devices kept in one place, he knew that the call might be important.

Maybe she's on call.

Andersen picked up her cell phone, expecting it to be the charge nurse begging for his wife to fill in for her.

Maybe it's that new nurse that filled in for her when she got sick. She dodged a bullet there, he thought as he cleared his voice to talk.

"This is Laura Andersen's phone. How may I direct your call?" Andersen had always playfully wondered what a job in stand-up comedy would be like. He wondered what else he would think of after two shots of whiskey.

His wife had been home with the kids that day and not in the eye of the storm. However, the male voice on the phone threw Andersen off right from the start.

"Lieutenant Andersen?" the voice asked.

Andersen's levity evaporated and his tipsiness was now gone.

"Yes," he responded cautiously.

Is this it? Is it who I think it is? How did he get this number?

"This is Alex Burns," the voice said as deafening silence followed.

Unbelievable.

Andersen's home phone, cell phone, pager, and work

phone had all been tapped for this very call, courtesy of nearly every federal and state law enforcement agency.

And somehow, Burns got my wife's cell phone!

"How did you get this number?" Andersen responded, shocked by Burns's ability to consistently do the unexpected.

They got Laura's phone number, he kept thinking.

"My colleague obtained the phone number at a party your wife attended," Burns calmly reported.

In the silence that followed, Andersen was at a complete loss as to what to say.

You were at a party that Laura was at? What?

He was stunned. Almost too stunned for words.

Is no place safe from these people? It's one thing to terrorize a nation and point a gun at the head of the US government, but his wife?

Angry, Andersen could muster a short response.

"You got some pretty big balls doing that."

Burns probably knew that Andersen's wife's cell phone would probably not be bugged or set up to trace.

"It's funny. When someone you love or care about is threatened, an average man or woman is capable of doing anything."

There was silence.

That statement had a great deal of meaning and hit the mark. Andersen immediately recalled David's response to being called a terrorist and his comment that it was the government "that had killed my wife" and threatened his friends and taken his life away. At the time, Andersen had an intellectual understanding of what David was feeling. Now with his own wife in the mix and her having contact with someone in their group, Andersen felt vulnerable. It was as if he was watching his wife innocently brushing up against the angel of death. Andersen was brought back by Burns's voice.

"I am not interested in harming you, your wife, or your children."

Children? He knows I have kids!

"You'd better stay away from—" Andersen started but was cut off.

"Shut up, Lieutenant. Here's the deal. I have more classified intelligence and data to sell off to enemy countries and to the highest bidder, but I won't. My team has pieces of the data and will be able to distribute this information in various forms and to various agencies ranging from newspapers to the Internet should we lose contact with one another. Other foreign governments, friendly and otherwise, will also be pulled into the mix if necessary. This plan has been carefully orchestrated to ensure that if anything bad happens to any one of us, it will be the Department of Defense's ass that will pay. All their secrets will be visible to the world. Because of the shit I have, the US will fall in the crosshairs of all enemy combatants as well as a few friendly governments. All this is arranged on a dead-man switch, and you'd all better hope for us to have long lives without complications. This crew is family to me. It's personal. If I sense I'm being followed or watched or if anything happens to them, I will rain hell down on everyone, and I won't be satisfied until the whole world is burning. It will be the Apocalypse like you've never seen. Am I clear?"

Andersen could easily hear the enmity in his voice, especially when it came to the last part. He also knew that Burns's question had been rhetorical.

I guess there's only one right answer.

"Yes."

What other response can I give?

"And to prove that I am sincere, I sent you an interesting summary of the events that happened from May 1 to May 4, 2011."

Before Burns could hang up and disappear, Andersen had to know one thing.

"Burns, wait. Why did you leave Caulfield with me to be interrogated? You knew he would tell me everything."

That had been a burning question that had been keeping Andersen up at night. He was sure that Burns's answer would make tactical and strategic sense.

"There were three unknown variables in the Merrimack Valley that needed to be either neutralized or contained. The FBI needed to be taken out of the mix - first distracted and then disabled.

You were one of the people who had the intelligence from Guantanamo, and who might have been able to make sense of all the pieces faster. And you might have been able to coordinate resistance to our plan and compromise the mission. Finally, there was Diane Welch, commandant of the state troopers who had tactical field experience and training in counterterrorism in Afghanistan, and especially the Swat Valley, Pakistan. She has some familiarity with my former bosses. Sadly, she is on leave."

"Sadly," Andersen thought.

Diane Welch was a colleague and a very good friend. She was a childhood friend from South Boston, best friend to his sister, and he had known her husband and family for years.

Was Burns sorry because he had wanted more of a challenge?

Andersen knew that she was on leave because her husband, Joe, was dying of cancer.

"'*Sadly,*' Burns?"

"No one should watch a loved one die."

Then there was a click, and Burns was gone.

Andersen quietly listened to the dial tone for a minute before he moved again to close out the line.

This is bad.

It was easy to see that Burns could feel others' pain and his own.

Leave me alone, and I might leave you alone. Mess with me and my loved ones, and I'll do what I said - I'll open the gates of hell, Andersen reasoned out as he sat quietly in his chair.

By the grace of God, casualties were limited to two hit men and one of his old colleagues, he thought. *It could have been a lot worse. It might be if we screw with him.*

Andersen turned over his wife's smart phone, and there was an image of three documents. They looked like "after action" reports of the events leading up to the capture of Oman Sharif Sudani and the events following it as well.

"'Captured?' He's supposed to be dead, killed on May 2," he said to himself as he got his glasses to read the documents attached.

The briefing prior to the planned attack was set for May 1, but it was delayed as a result of the two leaders — the logistics field agent, Burns, and the team leader, Anthony Maxwell. They had fundamentally different objectives: Burns's plan was to land, strike, kill, and take the body for evidence to show the world that when you mess with America, you will be hunted down, found, and killed.

I can agree with that, he thought as he tried to change the font to a larger size.

It was easy to see that Burns wanted to demonstrate American resolve, and even if you were the "untouchable" Oman Sharif Sudani, "we would find you and kill you."

Andersen viscerally agreed with the approach. Similar to outbursts, he wasn't a vengeful man; however, Sudani had killed a lot of people, and Andersen did believe his death was a good deterrence.

On the other hand, Maxwell had another plan: he proposed that they kill all the witnesses, abduct Sudani, and use his knowledge to destroy the terrorists' hierarchy, and their command and control center. While Maxwell's plan had the advantage of being a long-term solution to terrorism, it ran completely contrary to the policy of not cooperating with terrorists, especially one who was responsible for thousands of innocent deaths.

While the president of the United States had authorized the "kill order," this ran contrary to Maxwell's plan, which had been authorized by Chairman Eric Daniels.

Who the hell is this guy to countermand an Executive Order from the President of the United States? Talk about balls!

The plan the president approved was nuked when Burns's helicopter experienced "catastrophic mechanical failure," and crashed while en route to the objective.

It was initially thought that Burns was killed, as was his plan. On May 2, Maxwell captured Oman Sharif Sudani, killed the witnesses, which included his wives and children, and was promoted for his efforts.

Andersen stopped reading to fully digest what was being said.

Wow ... we let that bastard live, he thought.

Andersen found himself getting angry; a hot anger about being duped, an all-too-familiar feeling these days.

Let the killer go and kill the witnesses? Women and children? What the hell? Aren't we supposed to be the good guy!?

Andersen forced himself to read on. He didn't want to, but he had to.

The report went on to indicate that the American government could not be seen as an organization working with the "prince of terrorism" so they went public with the "kill plan." Everything would have been all right if Burns had not survived the helicopter crash. But he had.

The classified memorandum went on to confirm that Maxwell suspected that Burns had survived the crash because there was no body at the crash site. While there was not much time to do a thorough search, Maxwell was convinced Burns was alive and listed him as MIA. What happened after was only speculation. The memo is clear that Burns had escaped the helicopter wreckage. It is learned later via medical reports that Burns's injuries to his head seemed consistent with an IED, while the injuries to his hands and arms were probably a result of the crash. The only substantive data that had been confirmed was when Burns arrived at a Red Cross site.

Andersen took another moment to reflect on the report.

How the hell did you make it out there? he wondered.

Taking in a deep sigh, Andersen returned to the document. He readjusted the size of the picture to read it better. It wasn't until the Red Cross sent Burns home that the foreign intelligence agency caught wind of it and dispatched its own team. By then, though, there was too much of a paper trail to make Burns disappear. And at the time, Burns seemed to have lost his memory.

Initially, the report stated that Burns's recovery was critical because of his "ability to recall critical data relating to security protocols and a possible agency's breach." The report finally ended with clear conditions and mandates:

> "Should a full recovery become apparent, his condition may need to be monitored. If it is apparent that Burns is compromised, he will need to be terminated for national security. All those in contact with him will need to be assessed for next steps."

The document was signed by Chairman Daniels and co-signed for the man reportedly responsible for handling the details of the "Burns case," Director T. Webber.

Andersen could not believe that this was even possible, let alone written down somewhere in an official summary report. But then these classified files were never supposed to see the light of day.

"*Fuck!*" That was truly the only word that captured the severity of this secret.

If this ever gets out, the American public and its allies will be really pissed. Who can I trust with this knowledge?

Burns knew he had put Andersen in a bad position. The truth could get him killed.

For ten minutes, Andersen sat quietly as he listened to the kids and his wife preparing for bed. As Andersen sat with his wife's phone in his hand, he had an idea. He started looking for the icons that would show any recent pictures. When he found them, he started to flip through them until he found the pictures of her party on May 1. It only took him a few moments to see what he feared. In various pictures, there were many different people and other nurses, but there were at least five pictures that had one nurse — a very shapely woman with raven black hair whose face was either turned away or obscured. Laura had complained, Andersen remembered, that she had not gotten a "good picture" of the nice nurse, a Ms. Smith, who assisted her when she fell suddenly ill and then gave her a ride home. Andersen was now thoroughly sick. The angel of death had given his wife a ride home, and had been in his house a few hours before hell rained down on everyone.

They were in my house as my children and I slept, Andersen thought.

Andersen felt a great weight hang around his neck. He was no longer interested in his single-malt scotch.

Chapter 20

"Inter arma silent leges"
"In time of war, laws are silent"

Three Months Later – August 16

TWO MONTHS WENT BY before Andersen and Helms met again at his office in North Reading. The Fourth of July festivities had been ready to move ahead, and security had been at an all-time high. In light of what had happened three months ago on May 2, being on alert was not an act of paranoia but simply the new normal in the Merrimack Valley.

Andersen had told Cratty and Helms about the phone conversation he had with Burns and the clear message Burns conveyed. Andersen did not share with Cratty the document he had seen on his wife's phone, however, but he had shown it to Helms. Helms had taken the phone and ran it through the magic of forensics and confirmed the image was indeed real. The plan was for Andersen to keep his wife's phone at all times as it was probably the only direct line Burns and his team would have with someone with whom they could communicate in case "hell rained down." Laura had been fine with it. She had wanted a new phone anyway.

Helms had called earlier that day to "go out to lunch," which was code for "we need to talk." Once he was in Helms's car, it became clear that they were heading to a small town just down the road called Wakefield.

As they turned down a quiet neighborhood street, they saw plenty of well-maintained, split-level homes with some cottage-style residences thrown in to add variety to the planned neighborhood. However, it was the small ranch house with an attached garage that caught Andersen's attention. While the style of the home was different from the surrounding ones, it was the overgrown lawn with dead grass and dying, misshapen shrubs that really stuck out. While other homes had the maximum of two barrels of garbage and recycling out for pickup, this house had garbage bags of all shapes and sizes left out. There was also a large pile of furniture put out as well. The furniture was far from used and damaged; it was good furniture that had just been thrown out.

To add to the disheveled nature of the house, the mail had piled up, and unread newspapers had simply been cast aside. Flower boxes were all dead, and there were no signs of life. Helms knocked on the door as Andersen patiently waited for a response and wondered who the hell was living here. Helms hadn't told him in the car; Helms liked surprises.

Why do I encourage this? I should have simply stayed in my office until he told me what was going to happen.

Still, Andersen didn't mind surprises too much. After the third knock, the door flew open, and there stood Davis in sweats, a tank top stained with either sweat, beer, or both, and her .45 automatic weapon just out of visual range behind her.

"Well, well, well … to what do I owe the honor of your company," she asked. But rather than wait for a response or to invite them in, she simply walked back into her house.

Well, this is a surprise, he mused.

The first thing Andersen noticed was that the house smelled of sweat and heat, and the room was stuffy without air conditioning.

"Air conditioning out?" Helms asked.

"It works. I like it hot and dark. Helps me when I'm lifting," Davis responded.

While there was a chair, a radio, and a television in the living room, there was also a set of weights and a bench, two mats, two dartboards converted to throwing star and knife targets, and a heavy bag recklessly drilled into the ceiling. Andersen remembered the first dumpy apartment he had rented when he had joined the force and before he had gotten married. His apartment, however, had been significantly better than this poor excuse for a home.

Even my barracks in the service were way nicer.

"I like what you've done with the place — early gym era with a hint of self-pity," Andersen had to say as he continued looking around, taking in the sights, smells, and heat.

Early in the week, Helms told Andersen that Davis had been held out to dry for the breach and put on "indefinite administrative leave" pending the conclusion of the internal investigation. The conclusion of this type of investigation would mean capturing the ghost Burns. It had been two months of being on leave, but it didn't look like Davis was wasting away or catching up on her sleep.

She's obviously not working on keeping her house up with the Jones. I hope she's not in the market for dating. It would be a hard sell right now.

When Andersen saw her last, he was impressed with her physical prowess to go toe-to-toe with a guy like Burns. For all intents and purposes, Burns was the operations center's devil incarnate, and it had been Davis who had nearly beat him, both physically and tactically. She might have been the only person to come close to stopping him. However, she hadn't, and her bosses were making her an example. It was evident she was taking another approach.

So you're preparing for round two, I see.

There was no doubt in his mind that she was conditioning herself for a rematch. She was leaner now, and actually seemed taller. Her muscles were well-defined, and Andersen was positive there were weapons all over the house.

I pity the fool that breaks in here.

"What can I do for you, boys?" she asked as she put her handgun on the floor and got back on her bench to press what Andersen figured had to be more than her own weight and then some.

Jesus! Are you on steroids too or are you just that pissed?

It was Helms's show, so he took the lead as Andersen searched for a clean wall to lean on.

"I have a reconnaissance mission of sorts that only we three can be involved—"

Davis bolted right up from the bench and interrupted

frantically by putting a finger to her lips, making a *shh* sound indicating possible listening devices.

"Already checked three times in the last two days. Your boys pulled the bugs out three days ago, and my people swept the place already," Helms said.

Davis shook her head in either disgust or amazement.

"Great. I got two agencies looking at me."

"Actually three. The National Security Agency was here the first four weeks but figured they would let us handle it," Helms informed her.

"Great. I don't even have to leave home to get a date," Davis responded as she pressed a set with twenty reps.

"I wouldn't worry too much about getting a date," Andersen blurted out.

Davis looked at him in shock but then smiled.

"I don't know you well enough yet, but I'll make you pay for that one, as good as it was," she said as she got herself back into position for another set of chest presses.

"I am impressed, Davis," Helms continued. "You Navy types don't strike me as 'lean, mean, fighting machines.' Even if you were Office of Navy Research and Development."

"I'm expanding my boundaries," was Davis's short response.

After Davis did another quick set, Helms took out three folded papers and handed them to her.

"Maybe this will help with expanding your mind," he added.

Andersen knew it was a copy of the images that Burns had shown him months prior. He was not exactly sure what Helms was planning, but for whatever reason, he did trust the Marine.

Davis read the document. Then she went back to the beginning and read it again. Andersen was beginning to analyze Davis; he was sure he would be spending some time with her. He already knew she would dedicate herself to a job. As she read, she walked around in small circles as if she had energy she needed to burn off. He also noticed she was fiddling with a necklace.

It doesn't fit her. It's too delicate for her, he thought.

"You confirmed the authenticity of these images?" Davis asked. Helms nodded affirmatively.

"So that's why he was so pissed."

Davis sat down on the bench and took a drink of something in a water bottle.

Water? No. Protein drink more likely.

For a moment, Andersen thought he glimpsed either softness or understanding. Probably understanding, he deduced, as Davis did not seem the soft and warm type.

But then she has that necklace. A crucifix, no less, he thought.

Helms took the only chair in the room and started his briefing with a question.

"So Davis, if you were running the operations center and this breach occurred on your watch, what would you do?"

Without hesitation, Davis took another swig and answered.

"It did happen on my watch. But to answer the question directly, I would hunt them all down and kill them all. Nothing personal. Just business. I would want to make sure all the secrets and loose ends were taken care of. The fact that there are four of them probably in different parts of the world and there are dead-man switches on the critical data makes it more difficult ... so if something happens to one or all of them, the classified information gets out, which greatly complicates things."

"Two women, one child, a blind man, and a former good guy, not exactly an elite task force or a group of crazed terrorists," Andersen had to say.

Davis just stared at him as Helms pressed on.

"But would you just let them go and hope Burns keeps his word and everyone on his team stays in good health?"

"No, I would first locate them all. Watch them for a period of time. Assess their strengths, find patterns and then create a plan to either simply contain them or take them all out at the same time while getting all the information at the same time. That would be the plan. Difficult but that would be the best plan."

"Could Cratty do that?" Helms asked.

Davis seemed to laugh a little at first before she answered the question.

"No. She would need a team that she could trust, and after the past few weeks of hearings, I am sure she doesn't trust anyone now."

Andersen gave Davis a nonverbal cue to elaborate.

"Before all this shit went down, Cratty was designated to get a promotion to deputy director. Burns and his team ended that. She had some personal family matters that happened too, but it was the security breach that ended her shot. The director, Webber, was busted down to deputy director for being MIA when the Merrimack Valley burnt up. He is a vengeful, pissed-off, little bald man who will make it his business to get rid of Cratty, especially since I'm gone. He needs a whipping girl."

It looked as if Davis paused for a moment and reflected before she went back to the original question.

"But there are at least three other people who could have the natural skills and talent ... and if I had a team I trusted, I could get them too. Chairman Daniels will certainly supply all resources to bury Burns and his team. What are you getting at, Helms?" Davis asked.

Andersen could see that Helms wanted to carefully articulate his words because he needed Davis and she could smell bullshit from miles away.

"I know you think that Burns destroyed your career. You got in his way, and he neutralized you. I'm surprised he didn't kill you, but you were not his target. He minimized collateral damage and focused on saving his people. He could pull the trigger on us at any point but made it clear that if we leave him alone, they will live and let live. I think that he will live up to his side of the bargain, but I am convinced our team won't. The problem I have with all of this is why would we kill our own citizens for telling the truth after we destroy their lives?"

Well...for a Marine, you are pretty articulate while blunt at the same time.

Andersen understood exactly what Helms was saying. As much as he hated being played and used, he had a deep-seated admiration for Burns, David, and all of them. If the tables had been turned and he had been in Caulfield's or the Littleton sisters' position, Andersen would like to think he could do the same things to save his loved ones.

I think I'd want to but could I? Could I go off-grid for years and plan all of this, and then carry it out successfully? I'd like to think I could.

Andersen found himself obsessing about that dilemma often, but right now he was trying to focus on what Helms was saying.

"And another thing. What else has 'our team' done? What other secrets and dirty operations have we done? I fought in two theaters of war to protect the Constitution of the United States. And for what? To hunt down our own citizens?"

Davis was pacing again, though she had shifted from circles to back-and-forth movements. She was now moving from corner to corner of the room, a throwing knife in her hand.

"So what do you want me to do?" she finally asked.

Helms leaned forward on his chair and looked her right in the eyes.

"I want to find Burns and his team first and bring them in before our team makes its move. They want to make this all go away quietly. I want the truth. I want you, me, and Andersen to be in play to assist or deter. I want Burns to know we aren't all assholes and that there are at least three people he can trust to keep innocent people alive. If we can do this, maybe we can bring him in from the cold, get all the information back, clear him and his people, and set the record straight. You know ... do the right thing. From what you know about Burns before he went rogue, was he good?"

Davis answered without hesitation, yet both her gaze and voice seemed far away, distant. It was clear she had been reading as much as she could about him.

"He was the best. He was excellent at what he did and never failed a mission. What was also clear, though, was that he liked to leave a lot of bodies around in addition to his target. It was his calling card. The more bodies on the ground, the better he seemed to feel. A psych report I read a billion years ago pegged him as a sociopath. That's why I can't figure him out now. He's totally different," Davis said before her voice trailed off.

"Sounds to me like that head injury might have given our guy a heart or a sense of duty or honor or loyalty. It looks like he put his friends over his own needs and his country," Helms offered.

"Some could call that treason," Davis baited.

"What do you call it, Jill? Some guy, sociopath or not, who was trying to do the right thing and kill a bad guy and won't just follow orders to 'go along to get along,' and is seen as a problem? You know the guy. Burns wanted Sudani dead. Didn't we all? And for being a stand-up guy at that one moment, he is nearly killed, and everyone who tried to help him was attacked and killed. Is that treason? Or a guy who decides that his needs and wants are secondary to doing something right? With his skill, he could have gone off-grid forever, and left the Littleton sisters, the little girl and Caulfield on their own. He didn't. How come? He's now a concerned citizen who just wants to be left alone but he now has a family. We know your guys won't let this lie. Do you want to be ahead of it or just watch freedom die and innocents be sacrificed and labeled terrorists?"

Andersen became distracted.

Jill? Is that really her first name? he wondered.

Davis's pacing had slowed down as Helms made his point. She was standing still in the middle of her living-room-turned-gym. It was easy to see that she was weighing it all as she fiddled with her necklace again.

That necklace has to have meaning to her. I wonder if it calms her or is a burden. It does mean something, though, Andersen concluded.

"What about him?" Davis asked, pointing to Andersen.

"I wouldn't be here if I didn't agree. I may not know everything about Burns and Caulfield, but I know it would be better for all parties if cooler heads prevailed and we found him first. I know Helms, and I don't have a vested interest in Burns's death. Just the truth. What about you, Davis? Do you want him dead, or do you want the truth?"

Andersen had to ask the obvious question. It wouldn't have been the first time someone in law enforcement had wanted revenge. Davis had both the skill and motivation to do it too. He needed to know that he wouldn't be used again to kill Burns.

"For now, I want the truth. If Burns produces more of this shit," Davis said as she pointed to the classified documents, "then I am on his side. I am not just willing to hold hands and put flowers into gun barrels. If I get more proof, I will provide him with more support than he would imagine."

Andersen looked at Helms and asked, "Is that her way of saying she's in?"

"I think so. She's not very good with being clear and really sucks with her people skills."

"Does she at least shower and clean up good? Every time I see her, she looks like shit," Andersen asked.

"I hope so. She won't get the 'Good Housekeeping Seal' for her home and I am positive she can't cook," Helms responded as he looked around the sparse, smelly room.

"Hmm. So that leaves just her looks? Jesus!"

"Hey! I'm in the room guys. And I am also armed," Davis said.

For the first time, Andersen saw her smile.

Helms sat back in the chair and sat right up again. Something had clearly poked his butt. He pulled a full magazine for Davis's handgun out from his backside.

"Really, Davis? Really?" was the only thing Helms could say.

Chapter 21

"Caveat"
"Let him or her beware"

Seven Months Later – December 7

IT HAD BEEN SEVEN months since Burns had left Massachusetts. His conversation with Andersen had gone as well as one could expect. He knew it had to be nerve-wracking to have a person like him and his team involved in your personal life.

Maybe that thought will make them rethink finding us and keep them all at bay, he hoped.

But Burns never really believed his old bosses would leave him alone. Since he had left his team, he had been strategizing the next series of scenarios that he would need to implement to keep ahead of the flood waters building up just behind the dam. He wanted to keep David, Becky, and Emma out of the main task force and use them, if at all, solely as backup and research. That was why he wanted them to accelerate their plan to get out of Canada and go to Europe. He had plans for Sam, however, while it was clear that she had plans of her own, and she would not be sidelined.

After two weeks, Sam arrived at a designated meet, and then they were supposed to go their separates for another month, only to return later to "spend time together." She wanted nothing to do with that plan and was going to be with him now. She showed up with one backpack, a purse, and was dressed to "fit in" as she was always capable of doing when she had to. Burns attempted to argue with her, but she did that thing of agreeing with every word he said as she stepped closer to him, narrowing the gap between them, lips inches away from each other, while looking straight into his eyes. He gave up, and she had been with him ever since.

Burns felt the happiest when they were in bed together. It wasn't just the sex. It was something about how they talked to each other. Sometimes they would just read together or would just have dinner out, and lose track of time.

Out of habit, Burns monitored the *Merrimack Valley Times* newspaper for local chatter about a prior hot zone. He would check the local town events, schedules, death notices, police logs and personal ads. There was one personal ad that stood out, one that Burns knew was a message for him and his team:

> "ISO missing friend who loved *Rime of Ancient Mariner* like me last May. The craziness happened in the MV, and we never finished our conversation, and his friends can't be found. Did enjoy our last phone call though and hope to talk, even though family is angry about our relationship. Hope to hear back soon. You know the number. L. T. Andy."

"Not really subtle, is it?" Samantha said after she read the message.

"It's actually good. You would need to know Coleridge is the author of the *Rime* and make reference to our crew, the time, and actually sign it with L. T. Andy, meaning Lieutenant Andersen."

"So what does it mean? 'Hoping to talk, even though family is angry about our relationship' is a good thing?" Samantha asked.

Burns looked over at her to respond but then saw that she had just finished dyeing her hair. He wanted to say that she had overshot her goal of light brunette to raven black but decided it would be safer to smile instead.

Samantha suddenly looked up at him with her two distinctive eyes.

"What? What's wrong with my hair?" she said sharply.

"Nothing. Just looking at you. Kind of sensitive, aren't you?" he lied.

"Hmm," Samantha muttered as she walked back into the bathroom to finish drying her hair.

Danger narrowly averted...proceed at your own risk.

David did remind him that it was not a lie when your spouse asked if something looked good and it didn't but you said otherwise. But he did have a qualifier to the rule: "But never let your significant other go out in public if they might be embarrassed and you didn't try to stop it."

Burns had to assess if Samantha would be too embarrassed if her hair was too dark.

Hmm. There was one advantage about being alone. Relationships can be really difficult, he thought.

Burns turned his attention back to the ad, and was positive about what it meant. The code and its location was a field agent's method of communication with either individual contractors or operatives in deep cover.

"I think Andersen has found someone in my old shop that he can trust and has something he needs to let me know, even though his superiors would not agree and support him," Burns yelled to Samantha.

"Maybe it's a setup, and they want you to bite," she offered when she returned.

Burns appreciated Samantha's healthy dose of paranoia.

"No ... Andersen wouldn't want to risk harm to his family. Something must be brewing, and he wants to get to us first before something goes south," he concluded.

Samantha agreed with him.

Yup. It does make sense. I wonder what's going on. I bet Daniels is getting a task force together. I bet he's really pissed that I made him look bad. Would he risk destabilizing the US government for ego, though?

"I think I'll respond tomorrow. Can you let Becky know about this? I'm sure everything will be fine, but I want them 'on station' and ready to move if they have to," Burns said as he turned the page to see if there was anything he missed.

"Already covered. It's Emma's birthday in a couple of weeks, and I was going to contact her anyway."

He smiled at the notion that they would be planning a visit in three months.

Maybe it will be sooner than later, Burns thought.

Andersen couldn't believe he had placed a personal ad in the local paper. What he had written made sense, he guessed, but it really seemed too much like the Cold War, like something you would read in a post-World War II or Korean Conflict spy novel.

"I know you think it's silly, but if he's out there and he follows protocol, he will still monitor a recent hot zone for at least six to eight months, at which point he will see it and respond in about two days," Davis explained.

Nodding as he put the paper down, he glanced at her and was still marveling at the fact that Davis was actually wearing a dress.

Well, that's a first. She actually looks feminine rather than ready for combat. I guess spending time with Laura has had an effect. Before that, every time I saw her, she was either in a fight or ready for a fight, he reflected.

While no longer part of the foreign intelligence agency, and now a private contractor with the Federal Bureau of Investigation in Boston, she embraced her assignment of "research and follow-up of last spring's crisis in the Merrimack Valley," and was working directly with Deputy Director Helms with support from the North Reading Police Department.

As a matter of fact, I'm her boss. I think I'll break that news to her later, he smiled.

Davis had grown on him and it was now commonplace for her to come over for dinner at his wife's insistence. While initially it was awkward, he saw a more sensitive side of her; she seemed to enjoy being in a family situation. At first Andersen had to stop Laura's efforts for matchmaking once she got to know her but much to his surprise, Davis did not reject the idea outright. Still, as she seemed more like his kid sister, Darleen, it was a thought that Andersen really wanted out of his head.

Davis on a date? Nope! I just can't see it. Time to think of something else.

"Okay, Davis. We'll do it your way. But if I get some love-starved, literate, Coleridge-loving Generation X response, you will need to field that one," he warned.

"No worries. I have been interested in dating again, and this might meet two objectives at the same time," she countered.

What? Is dating a mission? She seems more like a guy at times than a woman. I swear she's got some serious issues going on.

"Is that how you think of dating? 'An objective?' How do you meet anyone?" Andersen asked.

"I only date to have sex—" Davis started.

Andersen shook his head.

You see! This is the very reason I have to stop Laura from matchmaking! Nothing good will come of this.

As if she had been reading his mind, Davis looked at him askance before she crossed her legs and brushed her dress of wrinkles.

"You know, Laura wants me to meet a nurse she knows that works third shift. It could work schedule-wise. We would see each other only in the afternoon, have coffee, sex, some conversation…" "Okay, Davis. That is way more intel than I need to know. I'll call you when I hear something. Now get out of here. I have to find some work to distract me from the thought of you having sex. With respect, it's like if my sister was talking about it and I just can't bear to hear it, thank you very much," Andersen said as he stood suddenly and moved to escort her out of his office.

He could see Davis smiling as she clearly allowed herself to be moved from his office.

"Helms was right. If I knew talking about me having sex would shut you up, I would have done that a while ago. Helms is brilliant for a former jarhead."

Jesus...those two are cut from the same cloth!

"Yeah ... he's a real genius," Andersen concluded and then shut his door with Davis on the other side.

Andersen turned and stopped as he was positive he heard Davis laughing outside his door until it started to recede.

God, she's a pain in the ass, he thought.

Standing in front of his desk, he thumbed through unopened letters as he listened to his voicemail on speaker.

While the majority was business, there was one that stood out.

"Hi Steve...I'm...ah...I'm sorry I've not been back in touch with you and Laura. The dog died about a month ago and I'm just trying to get things done in the house. Guess she couldn't live without him either. Say hello to Darleen and I'll talk to you soon."

Sitting down, he hit the option to listen again just to make sure that the emptiness in Diane Welch's voice wasn't something he was imagining.

Shit. She sounds awful, he thought as he imagined what he would be like if Laura had died suddenly of pancreatic cancer. Sitting in his chair, he truly wondered what would help him if he was in his good friend's shoes and the love of his life after three decades of being together died suddenly.

"Work," he said aloud as if he wanted to make sure he didn't forget what he said.

Andersen had been seriously thinking about having her assist in the plan to reengage Burns again. It made a lot of sense based on her skill set in terrorism and counterterrorism, not to mention her combat experience in Afghanistan. Also, Burns was clear that if she had been in the mix back in May, she would have been a force to contend with in regards to completing his mission.

Well, that's a ringing endorsement of her abilities if I ever heard one. And what the hell was Burns talking about? She was in Afghanistan. She never talked about being in Swat Valley, Pakistan... Unless he knows something I don't know, which he probably does.

Before hearing this message though, Andersen wanted her on his team, and Helms agreed it made sense to recruit her.

Picking up the phone, he put it back down before he even dialed.

She's not going to answer. I've left her a billion messages and stopped by her empty house. She's not there...

Sitting quietly for a few minutes, he tried to come up with a plan that would get her attention.

Swat Valley...Pakistan. Burns said she knew his old bosses...

After ten minutes of ruminating, an idea formed rapidly as he finally remembered two important names.

"That's it. Nine and Ice - Sgt. Thomas "Nine" Williams, and his Ojibwe friend, Lance Corporal Daniel "Ice" Maddox. If anyone can find her and bring her in, it's her men," he said as he smiled at his brilliance.

They may not have been under her command in years but they're tighter than families I know, he thought as he recalled at least five or more barbeques they had every year to remember the fallen.

"PFC Parks," he said aloud as he started his email and internet search for both men.

There was something about PFC Parks that was important. Died too young, I think. They have to know what Burns was talking about. Swat Valley, Pakistan, Webber and Daniels, he thought.

"They'll bring her in," he said as he found Nine's phone number.

Denise Cratty had spent the last couple of months getting used to the new operations center, code-named "Bravo," in New York City. The move was easier for her since her mother had passed away two months prior and her life partner was no longer interested in adopting. It had also become clear that her partner no longer wanted to be with her as well, which made leaving Massachusetts even easier.

Personally, Cratty was feeling in a bad place.

Professionally, she had been through hell and back with the internal investigations, reassignments, and overall restructuring of the operations centers. She had used much of her political clout defending her decisions, which included deploying Jill Davis to the auxiliary control room.

And what the hell were we supposed to do? We followed protocol and procedure. And at least we were there to deal with the situation and not incommunicado, she often thought as she would clean her apartment or be shooting at the range.

Cratty didn't believe for one minute that Davis had been a weak link in the chain. She had heard of the famous Alexander Burns, and was convinced that Davis had been the best person to have taken him on. Cratty never really knew if Davis was aware of how she had tried to get her reassigned to New York so that they could work together, but her attempt had been met with such a political firestorm that Davis opted to become a consultant for the FBI.

You might have made the right decision, Davis. The bullshit piles up so high around this cesspool you need a ladder to stay above it.

Cratty was surprised and touched that Davis sent her a serrated knife with a graphite handle with a note attached saying, "Thank you."

Not exactly a nice bracelet or a pair of earrings, but a nice gesture, she thought at the time.

Cratty's immediate boss, Deputy Director Thomas "Steel" Webber, ran the operations in Maryland, code-named "Alpha."

Not very original names, Alpha, Bravo...but then what did you expect from a guy like Webber, she thought as she eyed her gift from Davis in her desk drawer. Right beside it were two small bottles of scotch that her team had given her as a going away present. Feeling thirsty, she moved the bottles to see if one of them was actually already gone, before pushing them towards the back of the drawer.

I got to get these out of here, she thought.

Still, her mind shifted to politics and wondered how it became so commonplace in an armed law enforcement agency. It was easy for her to see that Webber was both political and ambitious. More importantly, he was not happy that this ghost Alex Burns and his homegrown terrorists were able to "walk away with classified data" from his newly acquired and transformed shop.

Hey, Webber! News flash - no one's happy about what Alex Burns did! You'd think you were the only one affected by this.

Burns had cost Cratty a possible promotion and Webber had been demoted from director to deputy director for being MIA when he had been needed most.

Still, Cratty was not remotely upset about not getting a promotion and especially not concerned about Webber's career. She was angrier that she was not given critical information about Burns at her level as a manager until after the breach.

Why the hell wouldn't they let everyone, supervisor and above, know about Burns? she thought.

When she read only a few samples of Burns's early missions and his profile, she was able to understand how he did what he did, but she was not sure why or why his mode of operation changed from not caring about collateral damage to minimizing it to an incredible degree.

Three fatalities? No innocent bystanders hurt or killed? That's not the Burns we know in the files. He's a different man now. He's a real different kind of problem now, she concluded.

As much as she personally wanted to get Burns, none of the data made any sense. Cratty had spent weeks trying to convince Webber and Chairman Daniels that, based on the information provided by their own analysts, Andersen's reports, Helms's forensic and behavioral science evaluations, and the leverage Burns possessed, the best course of action would be to "let sleeping dogs lie."

At the most, Cratty was all for locating them but only for observation and not for detention or termination.

Keep them under surveillance, contained and bottled up. Not too difficult and to be expected by a guy like Burns. But to go on the offensive with the shit he has? Dumb.

But Chairman Daniels was petitioning the Pentagon, Homeland Security, and even the West Wing Executive Office to pursue this group to "bring them to justice."

So earlier that day when Cratty signed for a sealed letter that was personally couriered from Alpha, she was pretty sure it would be a memorandum with an order she thought would lead to hell. Cratty opened the letter, found her glasses, and read it:

To: T. Webber, Deputy Director, Op. Alpha Center
Cc: D. Cratty, Manager, Op. Bravo Center, 2nd Shift
J. Glenn, Manager, Op. Bravo Center, 1st Shift
From: Chairman Eric Daniels
Re: Operation Rising Phoenix (Classified)

As discussed, you are to assemble three separate teams that are fully resourced and equipped with material, intelligence, and support staff to locate, apprehend, and bring to justice Alexander J. Burns for his role in the May 2 attack in the Merrimack Valley, Massachusetts, the dispersion of explosives and threat to detonate, the cyber-attack on two federal agencies and local government offices, attack of federal officers, the abduction and accomplice to the murder of a federal agent, the stealing of classified documents, espionage, and treason.

In addition to Mr. Burns, the following two people are persons of interest who are to be apprehended, detained, questioned, and if necessary, prosecuted for their role in assisting Mr. Burns: David Caulfield and Samantha Littleton.

Rebecca Littleton (sister of Samantha Littleton) is a person of interest in her role of possibly assisting the abovementioned names in the formation and commission of the abovementioned charges.

This order is to take effect immediately with daily updates scheduled for 0900 hours. As your subordinate, Ms. Denise Cratty, has made it clear that she disagrees with this course of action, I am assigning you a new manager, Mr. Jeffery Glenn, to be lead on operation center's Bravo team, NY. He is to lead the first shift. I want Ms. Cratty to run the second shift. Also, Ms. Cratty's performance review is bumped up by six months.

I will not approve any further time off for senior staff until this operation is launched and successfully completed.

E. Daniels

Cratty put her glasses and memorandum down on the desk, and turned to get up.

Jeffery Glenn? He's going to run this operation? He's got some skills, but they aren't up to Burns's level, she thought as she stretched her legs.

He's got some shit going on too. I guess with Maxwell gone and Webber not wanting to get his hands dirty, I guess he drew the last lot and lost, she pondered as she finally felt circulation returning to her feet.

Taking in a deep breath, she walked to the Plexiglas that looked over the operations center's main control room. Peering to see where her new staff were stationed, she caught a glance of her reflection, which was a lot thinner and clad in much darker clothes than she usually wore.

She had a momentary flashback to seventh grade with Sister O'Neill standing over her as she carefully recited part of Dante's work about hell.

"All hope abandon ye who enter," she recited in her head.

Dante? The Inferno? Was that it? she wondered.

She stood silently peering into the dark center and wondered if she would really have enough time in her role as manager to appreciate her new office and operations center. Silent and still, she couldn't stop staring through the transparent glass as she kept repeating the same thought that this course of action would pave the way to hell.

"Nothing good will come of this," she said to herself.

Becky had finished reading her evening e-mails and texts from Samantha, and she had to admit that she could see that her sister was a changed woman. She spoke about wine, shoes, movies that Burns and she had seen, places they were going to, all things that Becky had always wanted her little sister to experience.

Finally, she found the right guy. An international spy, domestic terrorist, fugitive ... but a nice guy, she thought.

Closing her laptop, she walked through the rustic, sparse hallway and stopped off at the bathroom to check the scale before she made her way to the kitchen.

Carefully stepping on the new weight scale, she made sure to take off her bracelets as if it might make a big difference. Closing her eyes at first, she finally looked down upon hearing the scale finally stop and saw the number.

"143 pounds...excellent!"

Stepping off the scale quickly, she waited until it stopped moving before she got back on to double-check. After confirming she was still 143 pounds, she smiled as she put her bracelets back on, and marched to the kitchen to make dinner.

Upon entering, she looked out the windows that gave her a grand view of the trees – three to four olive trees flanked by pines mostly surrounded by carob and pomegranate, a few small homes, and a clear though narrow view of the Mediterranean Sea. She found that every time she entered her small kitchen, she was still struck by its view, so different from how she lived as a child.

I just can't believe it.

"Grilled chicken with tomatoes, cucumbers and feta cheese with pastitsio, olives, yogurt and figs...God I love this place," she said aloud as she pulled together the ingredients for late dinner.

Late by American standards.

On Kea, dinner at 8:00 p.m. was early. It was so different living on a Greek island, more different than she ever imagined.

Several months ago, she remembered shedding one identity for another under the shadow of a national crisis. A crisis she had helped to create.

Now how did we make it? That was just crazy! What the hell were we thinking?

Becky remembered having a glass of wine on St. Catherine's Street in Montreal on May 2 several months ago. But it was so surreal that she could not relax. Even her usual rum and Coke was not doing it for her.

Having entered Canada as Canadian citizens, they were already set up to live not far from McGill University, which would have been great. But the plan was accelerated, and the "final port of call" came up four months earlier than planned. By the Fourth of July, David, Emma, and Becky were now the rich Canadian expatriates living in a modest home overlooking the Mediterranean Sea on the island of Kea.

Wow. Amazing what money and forged passports can do, she had thought.

When she moved to Kea, she wasn't sure if she would like it, but when she saw her new house, she immediately fell in love with it. Their home was rustic, clean and orderly. The ocean breeze was constantly making even the hottest days cool. Growing up, Becky and Samantha never had money. Living a lifestyle where money was not a problem, and food and shelter were never in question, was a little strange to her in the beginning. But after awhile, Becky truly was thrilled to have the resources to provide Emma the attention and stability in life that she herself never had.

At least she'll have a better life than we did. I bet Tony would have loved this. He sure would have been happy that Emma was here.

Remembering back to her youth, Becky had always thought that all the Mediterranean islands were constantly filled with visitors, vacationers, and tourists. That was reinforced by all the brochures she used to look at when she lived back in the States ... before Emma, David, and Burns. Tourism on this Greek island was nonexistent, leaving only the natives and next to no non-Greek visitors.

The tourists have no idea of this treasure, she thought as she pulled together her new, favorite cooking pan.

Even though it was clear to Becky that their new neighbors saw them as *xenos*, it didn't take very long for Emma to win the hearts and minds of their children and grandparents.

Becky originally thought it had been David's idea to live on Kea, but she found out late in the process that it had been Burns who had chosen the small Greek island. Choosing to live on Kea was not based solely on its beautiful location and perfect climate; it was also small, and newcomers stood out. Plus, everyone watched out for everyone else. If there was someone new in town who was not a tourist, David and Becky would hear about it first.

Always thinking tactically, aren't you, Burns? Becky thought to herself.

Reluctantly, she found herself agreeing with him.

This is what you do. Do you have any hobbies or something? Are you always thinking in terms of next steps? I guess we all have our reasons to be hyper-vigilant, she thought.

Becky did know that being on Kea was bittersweet for David. She knew he had always planned on going to one of the Greek Isles with Jenny, his first wife. Kos was the original place they had talked about, but they had never gotten the chance. David had never planned on her dying; he had never seen any of this at all.

None of us did.

For a long time, Becky thought that while David cared

for her and Emma, she also knew that living on Kos would have seemed wrong to him in memory of his deceased wife. A month after their arrival, Becky noticed that he seemed more depressed. Having spent a great deal of her life looking at her own various mental health symptoms, she knew firsthand the classic signs — not sleeping, not eating, unplanned weight loss, and irritability. So when she confronted him about what was wrong, she expected him to tell her that he missed his wife and wanted to be with her. Becky knew she was strong, but she was sure that she would just die if he left.

She had very few relationships when she was younger because she always had obligations — obligations to her parents, Tony, and Samantha. The guys who were drawn to her were always immature and not responsible. They were nothing like her brother, Tony, and definitely nothing like David. He had never made any commitments to stay with her, but it was clear that he was committed to be Emma's parent. Becky assumed that was the only reason why they were together.

Still, when she confronted David about what was wrong, and prepared herself to be hurt, his response surprised. When he told her he was anxious about the future and feared losing her and Emma and not being able to live with that, she was just speechless. Becky felt her heart break as he revealed to her his greatest fear.

He can't live without us. He can't live without me.

She could only rush to him and hold him as she tried not to burst into tears.

How can he be so strong but so vulnerable? she thought.

She thought she had never seen a man like him until she took her time to think about it, and came up with another.

Maybe it's the same thing with Sam and Burns, she wondered.

Shaking out of her daydreaming as she finished cutting her tomatoes, she was drawn to hearing Emma and David doing their evening ritual of dancing to music. Emma would choose music, and they would dance together. Becky knew that Emma wanted to live the way they used to in Rhode Island because Aunt Samantha lived with them and she would see Uncle Alex more often too. Emma also liked the snow as well, but she really liked to swim and dance. Becky was always impressed with her choice of music. Today, she chose "Uncle Alex's song," Johnny Rivers's 1966 classic "Secret Agent Man." The irony was not lost on her. David and Emma had just finished when Becky announced her presence.

"So how long were you watching?" David asked while he rubbed sweat from his forehead.

"Not long. You're both totally cute," she had to say. But she had more to say once she told Emma to get ready for a trip to town for an early lunch.

"Sam just let me know that your Lieutenant Andersen needs to communicate with Alex. Sam doesn't trust him or the situation, but Alex is seldom wrong. Just to be careful though, Alex wants us to be 'on heightened alert.' Sam also sent a couple of places, people, and events she thinks we should research and possible relocation ideas for Sam and Alex. They both want to be closer to us."

Becky watched David produce a small smile.

"That would be very nice. I never thought I would say this, but I miss the Rhode Island days just like Emma. We were all together," he commented.

Becky came up close and embraced him. As always he held her too. It was funny that David always wanted to be held by her but rarely initiated sex, as did she. But neither took it personally as they were very intimate in other ways, and they loved each other. Becky knew that.

I believe it, she thought often.

Becky started to break away when she heard Emma giggling. Emma always giggled when the adults "huggied." Unlike Samantha, David didn't pull away from her hugs. He never did.

I don't think I could ever live without you, she thought as she closed her eyes in his embrace.

David woke with a start. He had been sitting out on the veranda when he took a rest from listening to his audiobook.

I must have fallen asleep again, he thought as he adjusted himself in his porch chair.

David could feel the sun on his face, but it was low in the horizon. He moved his head slightly and was able to hear Emma laughing as he heard pots banging in the kitchen. Sound was his new, primary sense, followed by smell.

It took years for him to learn how to navigate the world in darkness. David remembered shortly after his wife died how he had never wanted to see again or even live, for that matter. Emma was the first to give him hope and a will to live. Once it was clear to him that she needed protection, he was able to motivate and mobilize himself. There had been many times over the years when David had second-guessed his abilities to expand beyond himself, but he always had t goal of keeping Emma and Becky safe. Of late, though, he found himself re-experiencing past pain. He had wondered if it was a sort of survivor's guilt.

Post-traumatic stress reaction? Unresolved grief reaction? Moderate depression? he would often think to himself.

It took years to rebuild his life with a focus of striking back and escaping. With that done and nothing to distract him, David now had time to just think. He used to be a psychologist who helped people, and who was married to a wonderful woman and had a beautiful life.

And in a flash, it was gone. All gone.

David felt for his audiobook and found it by the side of the chair. Over the past several weeks, he had had difficulty sleeping, and his appetite had waned. He typically was able to exercise, and that made him feel better; however, he kept focusing on "preparing for the future."

And what future is that now? Now what am I preparing for?

Because they all slept in the same room, David found that he felt better when he was awake at night, listening to them sleep, and then he would sleep during the day when he could easily hear them milling about. David knew by Becky's comments that she had noticed that his clothes were loose and that he was getting up at night.

One of the things he loved about Becky was her bluntness. It didn't take her long to ask what was wrong.

"e hell is wrong with you?" she asked him find yet another text-to-speech article on

ng "nothing" or making something up like

he used to in his prior life, he articulated what he had been thinking for weeks.

"I thought when we had gotten our leverage and escaped, we would be free. You know, find a Greek island, and just sit back and meditate all day. I thought if we could do the impossible, pulling off this mission and getting away with it, I would finally be able to relax. And the fact that no one got killed was just miraculous."

David immediately thought of Anthony Maxwell and felt guilt. *There was his death ... and the two others. They count.*

David knew he hadn't pulled the trigger, but he also knew he had helped set it up.

A time to pluck up that which is planted, a time to kill, and a time to heal, he thought to himself.

David pushed the random thoughts out of his head and continued before he forgot his point.

"But I find myself more anxious now than I ever was back in Rhode Island and Boston. I feel like I'm on borrowed time, like we all are on borrowed time. And I am not sure I could survive without you and Emma. I can't lose you. I can't go through that again, that loss."

David felt quiet as his voice felt tired and cracked toward the end. He felt a great burden lift from his chest as he revealed his greatest fear to the woman he loved.

He heard Becky sniffle and her chair move as she came over to hold him tightly.

For the moment, David felt at peace in her arms.

Presently, David's rumination was interrupted by Emma's and Becky's laughter, making him instinctively smile.

David found his book's cue button, and the last stanza

came to life in the form of a low, professionally trained voice:

> "...*Thus play I in one person many people,*
> *And none contented: sometimes am I king;*
> *Then treasons make me wish myself a beggar,*
> *And so I am: then crushing penury*
> *Persuades me I was better when a king;*
> *Then am I king'd again—* ..."

David turned it off immediately as he knew the rest well. He found himself listening to more poetry similar to Shakespeare's *Richard II*.

No wonder I'm depressed. Anyone listening to this stuff would be depressed, he thought.

Then David made a curious connection.

Maybe I'm trying to go back to someone I used to be? But now I am a father and a fugitive ... kind of a strange mix, he thought.

David was about to make a note in his audiobook's digital recorder to order more comedies and drop the orders for the serious plays when he heard Becky call him from the house.

"Dinner!" Becky yelled.

Well...maybe later, he thought as he pushed himself out of the chair.

"And Emma, move your shoes! I'm falling all over them. Why do you always have to leave them in my way?" Becky continued.

"Sorry," came Emma's singsong reply.

"Good thing she's cute," he said to himself as he made his way to the kitchen.

Upon entering, he straightened his back while taking in a deep, cleansing breath of sea air. The fading warm rays of sun still felt good on his face.

Tomorrow will be a better day, David thought.

Epilogue

"Bene qui latuit, bene vixit"
"One who lives well, lives unnoticed,"

– Ovid

March 14 – Ten Months after The Merrimack Valley Crisis, May 2

DIANE WELCH STOOD QUIETLY in front of her beloved's newly placed headstone as the sun's rays fell on her back. With no tears left, she stood "at ease" as she spoke to her husband the way she always did.

"I know it's late but I'm not ready to stop talking to you. You are so good at listening," she said quietly.

"Well, I already told you the kids are doing well and everyone's healthy but they're still worried about me, though. They were definitely happy about me consulting to the FBI. Well, more like observing than consulting. Steve wants me to be on a special task force to get the crew that screwed up everything last May. I know. It was crazy back then, I guess," she said.

"My former CO, John Helms, is the Director and I guess he's taking lead of bringing these people in. You remember him. He's the guy that helped with me getting home and keeping the vultures off of me. Good guy..."

Welch tilted her head down to make sure the inscriptions were right. The sun felt good on her neck, and the smell of pine was still strong in the air.

With her voice trailing off, she searched her memory to make sure she didn't forget anything.

I know there's something else, but what?

"I told you about me staying here and not going to Florida, right? I can't do that. Death's waiting room with nothing to do but look at sunsets and play shuffleboard. I'm glad we never considered living there. Maine's coast was a much better idea. Anyway, there's no way I'll leave you now. I am getting rid of the furniture we talked about donating and the kids have all their stuff, too..."

Welch first heard and then saw a vehicle slowly coming to a stop just in her line of sight. Looking at the jeep, she smiled as she knew exactly who it was.

"You know, just when you think the boys are nags about things, you realize that former Marine staff are worse. I mean 'Marine' staff. I know there's no such thing as 'former' Marines."

Watching two ostensibly large black men exiting the jeep, Welch's smile broadened as the jeep seemed dwarfed by their size once they were beside it.

"Well, Joe, I gotta go. Looks like Ice and Nine are here to make sure I'm fit to go to Boston tomorrow. I'd love to give them shit about it but I have to say that Nine's right – I would have found a lame excuse not to go. I have to give Steve shit for unleashing these two bloodhounds on me," she said as she watched both men casually leaning against the jeep.

Well, "casually" is not the right word. More like "unobtrusively scanning the area for line of fire," she thought as she kissed her hand and then planted it on the stone.

"Love you, Joe. Talk to you soon."

Walking down the slight incline, Welch found herself feeling sad as she always did when leaving him. While the loss was getting a little bit better, it reminded her of so many others. Still, she did feel a little better, hopeful, to have work tomorrow.

Well, I hope it's an easy day. Jumping right back into the shit is really not on my to-do list. But still...it might make the day go faster if something exciting happened, she thought as she approached her former team.

List of Characters

Alexander J. Burns – aka "**Falcon**." First seen in *Albatross*, Burns survives a helicopter crash while en route to a black-ops mission to kill the terrorist leader, Oman Sharif Sudani. Brain injured, he gets treatment that helps him regain his memory, and realizes he is a logistics field operative for the Foreign Intelligence Agency, FIA.

Samantha Littleton – aka "**Raven**." Introduced in *Albatross* and carries into *Raven*. She is the first to find Burns being sedated in the hospital and facilitates his transfer to an outside psychologist who specializes in assisting trauma victims regain their memories.

Dr. David Caulfield – aka "**Samuel Coleridge**." First seen in *Albatross* and carried through to *Eagle*, he is the psychologist who treats Burns and helps him regain his memories. He is also the one who convinces his friends to take the fight to the Eric Daniels' organization, the FIA.

Eric I. Daniels – aka "**Eagle**." Mentioned in *Albatross*, first seen in *Raven*, and fully elaborated on in *Eagle*, he is the

Chairman of the FIA, a privately held, clandestine, intelligence agency that will work on behalf of the United States government when it aligns with its own interests and objectives.

Becky Littleton – aka "**Tiny**." Introduced in *Albatross* as Samantha's older foster sister as well as Emma Littleton's primary caretaker after her brother, Tony, is killed by the mob. In *Raven*, her lethal skills grow exponentially.

Steve Andersen – Lieutenant, North Reading Police Department, MA, is the first to interview Dr. Caulfield, aka "Samuel Coleridge" in *Albatross*. Additionally, he was on the team of Army Intelligence at Guantanamo that connected the dots to locate Oman Sharif Sudani. He is also best friends with Diane Welch who grew up in the same South Boston neighborhood.

John Helms – FBI Director, Boston Regional Office, he is the first to detect the diversions and covert plan in *Albatross*, as well as attempting a negotiation for peace with Burns in *Raven*. He was also Diane Welch's CO briefly in Afghanistan.

Diane Welch – Commandant, Massachusetts State Troopers, first mentioned and then appears at the end of *Albatross* as Steve Andersen's close friend. She is mentioned by Burns as a person of interest, a key player targeted for deceiving if she had been around during the crisis.

Thomas "Steel" Webber – First seen in *Albatross* as the "face" of the FIA, and Eric Daniels' top man. He's the director of the agency, and was the team leader responsible for keeping Burns sedated and monitored.

Jillian T. Davis – aka **"Cougar."** First seen in *Albatross*, she is an off-duty manager of the FIA's Operations Center who is recruited to courier top secret, external hard drives to a secure location.

Denise Cratty – Introduced in *Albatross*, she is the on-duty manager of the FIA's Operations Center when it is compromised.

Emma Littleton – Introduced as a baby in *Albatross*, she is under the care of Becky Littleton and David Caulfield. Becky's brother, Tony, was Emma's original caretaker before he was killed.

ALSO BY J.M. Erickson

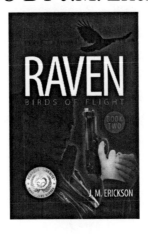

Raven

2013 READERS FAVORITE INTERNATIONAL BOOK AWARD WINNER
Honorable Mention in Fiction - Action

Alexander J. Burns, former counter-terrorist specialist and his partner Samantha Littleton are in hiding. After years of planning, Burns' team of civilians pulled off a shocking maneuver: they breached government security, stole top-secret documents, and then disappeared. In the process, Burns released a destructive computer virus known as "Albatross" to millions, infiltrating the security of three federal agencies, and immobilizing all levels of law enforcement, bringing the entire Northeast region to a standstill. Burns holds on to those documents for the security of himself and his team; they're the price of their lives, and their safety. But now, the FBI has approached him to retrieve those documents and negotiate for Burns and Littleton to come in from the cold. Although Burns' friends are thousands of miles away, suddenly they are threatened as well. With his friends under siege and the woman he loves targeted for death, it's time for Burns to go back to his old way of life, with killing and destruction all in a day's work. As new alliances are formed and unexpected enemies emerge, Burns has to decide whether he is willing to make the ultimate sacrifice. Following the success of the critically acclaimed Albatross, the Birds of Flight series continues its tradition of fast-paced, character-driven psychological thrillers with the gripping page-turner Raven.

Learn more at: www.outskirtspress.com/ravenbirdsofflight